A SAS® Companion for Nonparametric Statistics

Scott J. Richter
University of North Carolina at Greensboro

James J. Higgins
Kansas State University

THOMSON

BROOKS/COLE

Australia • Canada • Mexico • Singapore • Spain
United Kingdom • United States

Acquisitions Editor: Carolyn Crockett
Assistant Editor: Ann Day
Editorial Assistant: Daniel Geller
Advertising Project Manager: Nathaniel Bergson-Michelson
Project Manager, Editorial Production: Belinda Krohmer

Print/Media Buyer: Rebecca Cross
Permissions Editor: Stephanie Lee
Cover Printer: Malloy Incorporated
Cover Design: Ark Stein
Printer: Malloy Incorporated

Printed in the United States of America.
1 2 3 4 5 6 7 08 07 06 05

For more information about our products, contact us at:
Thomson Learning Academic Resource Center
1-800-423-0563
For permission to use material from this text, contact us by:
Phone: 1-800-730-2214
Fax: 1-800-730-2215
Web: http://www.thomsonrights.com

Brooks/Cole—Thomson Learning
10 Davis Drive
Belmont, CA 94002
USA

Asia
Thomson Learning
5 Shenton Way #01-01
UIC Building
Singapore 068808

Australia/New Zealand
Thomson Learning
102 Dodds Street
Southbank, Victoria 3006
Australia

Canada
Nelson
1120 Birchmount Road
Toronto, Ontario M1K 5G4
Canada

Europe/Middle East/Africa
Thomson Learning
High Holborn House
50/51 Bedford Row
London WC1R 4LR
United Kingdom

ISBN 0-534-42220-9

CONTENTS

Preface

This companion is designed for anyone who desires a guide to using SAS to carry out nonparametric analyses. It can serve as a lab manual for students enrolled in a course covering nonparametric methods where SAS is used for computing. It may also serve as a useful reference for data analysts who use SAS and wish to learn about the capabilities of SAS for performing nonparametric analyses.

Nonparametric methods were once regarded as quick, hand-calculation methods suitable only for simple designs and small sample sizes. The availability of high-speed computing to carry out computer-intensive methods has changed that. In addition to traditional methods based on ranks, computationally intensive methods such as permutation tests, exact tests for contingency tables, bootstrap estimation, curve smoothing, and other robust methods have become an essential part of the data analyst's tool kit. As a result, statistical software packages have become indispensable for implementing nonparametric procedures.

SAS is probably the most widely used software for statistical analysis. Many procedures for carrying out nonparametric analyses have been incorporated into SAS software, and in recent versions computationally intensive methods have been added. We assume that readers have some familiarity with SAS and will be able to execute simple SAS programs. Generally, only analyses that are incorporated into existing SAS procedures are presented, although in several cases we give references that the more advanced SAS user may find useful.

An important feature of this companion is that all of the SAS examples presented are self-contained in the sense that they can be entered into SAS as they appear and then be executed. Thus, the user does not have to deal with issues of creating SAS data sets before using the programs. In addition to presenting the "how-to" of obtaining nonparametric analyses in SAS, we also give brief introductions to the methods illustrated and indicate how SAS calculates the results it presents. We especially try to point out where SAS does things that might be different from what is typically presented in textbooks.

In selecting topics, we started by considering data analysis problems that commonly arise in practice. In most cases, SAS has nonparametric procedures for solving these problems. However, for a few situations, this is not the case, and in these situations we have tried to point the reader to alternative options such as use of the SAS Macro language when they exist. Although there are some nonparametric computations SAS will perform that are not documented in this companion, we feel that we have presented the most widely used methods available in SAS. Most of the programs were written using the capabilities in SAS Version 8 (SAS Institute Inc., 1999), but certain results in Chapter 2 require Version 8.2 or later, and the methods of the last section of Chapter 9 require SAS Version 9 (SAS Institute Inc., 2004).

1
One Sample Methods

1.1 Binomial Test for the Median

Suppose that we have a random sample from a population that has a continuous cumulative distribution function (cdf) $F(x)$, and let $\theta_{0.5}$ denote the median of the population. We wish to test

$H_0 : \theta_{0.5} = \theta_H$ against a two-sided alternative. Tests of hypotheses for medians are typically used in

the same situations that are appropriate for tests of hypothesis for means.

Let $X_1, X_2, \ldots, X_n$ denote a random sample from the population. Let B denote the number of X_i's out of n that fall above the hypothesized median θ_H. If H_0 is true, then each X_i has probability 0.5 of falling above θ_H, so B has a binomial distribution with probability $p = 0.5$. If the true median is other than θ_H, then B has a binomial distribution with probability $p \neq 0.5$. Thus, the test of hypothesis for the median is equivalent to testing $H_0 : p = 0.5$ versus $H_a : p \neq 0.5$. The p-value is determined from the binomial distribution with $p = 0.5$, and is the probability of obtaining a value of B equal to or more extreme than B_{obs} or $n - B_{obs}$, where B_{obs} is the observed value of B.

1.2 Confidence Interval for the pth Quantile

Now consider the problem of making a confidence interval for the pth quantile θ_p, the point at which $F(\theta_p) = p$. Place the observations in order to obtain the *order statistics*, $X_{(1)} < X_{(2)} < \cdots < X_{(n)}$. Confidence intervals are of the form

$$X_{(a)} < \theta_p < X_{(b)}.$$

This inequality holds if and only if at least a and at most $b - 1$ of the X_i's are less than θ_p. Thus, for a $100(1 - \alpha)\%$ confidence interval, a and b are chosen to satisfy

$$\sum_{x=a}^{b-1} \binom{n}{x} p^x (1-p)^{n-x} = 1 - \alpha$$

In practice, the values of a and b are chosen to make the probability as close to but without going below $1 - \alpha$.

SAS Implementation

The **UNIVARIATE** procedure can generate an exact test for the median of a population, as well as confidence intervals for the median and other percentiles.

Example 1.1. The data in Table 1.1 represent the sodium content in milligrams for 40 packages of a food product. The intended value for the sodium content is 75 mg.

Table 1.1. Sodium content of 40 packages of a food product.
72.1 72.8 72.9 73.3 73.3 73.3 73.9 74.0 74.2 74.2 74.3 74.6 74.7 75.0 75.1 75.1 75.2 75.3 75.3 75.3 75.4 76.1 76.5 76.5 76.6 76.9 77.1 77.2 77.4 77.4 77.7 78.0 78.3 78.6 78.8 78.9 79.7 80.3 80.5 81.0

The following code requests a two-sided *p*-value to test the hypothesis that the population median equals 75, and also produces 95% confidence intervals for the median and selected percentiles. The **cipctldf** option specifies that confidence intervals be printed, and the **alpha=0.05** option specifies that they have 95% confidence (the default confidence level is 95% if the **alpha=***level* option is omitted).

```
data ta1_1;
input sodium @@;
datalines;
72.1 72.8 72.9 73.3 73.3 73.3 73.9 74.0 74.2 74.2
74.3 74.6 74.7 75.0 75.1 75.1 75.2 75.3 75.3 75.3
75.4 76.1 76.5 76.5 76.6 76.9 77.1 77.2 77.4 77.4
77.7 78.0 78.3 78.6 78.8 78.9 79.7 80.3 80.5 81.0
;
proc univariate data=ta1_1
       mu0=75      /* Specifies population mean value to test */
       cipctldf    /* Requests confidence intervals for percentiles */
       alpha=0.05; /* Specifies alpha level */
var sodium;
run;
```

The two-sided *p*-value for the test that the median is equal to 75 is 0.0533 **(1)** and is based on the sign test (see Chapter 4). The statistic $M = [(\# \text{ values} > \theta_H) - (\# \text{ values} < \theta_H)]/2 = B - n/2$ used by SAS is equivalent to B described in Section 1.1. An approximate 95% confidence interval for the median is (75.0, 77.1) **(2)**. The last three columns under the "Quantiles" section of the output indicate that 75.0 and 77.1 correspond to the 14th and 27th largest values in the sample and give the exact confidence level, in this case 96.15%. The values for the endpoints of the interval are chosen so that the exact confidence level is at least 95%. Confidence intervals for the quartiles and other selected percentiles are also given. Note that for extreme percentiles such as 1%, 5%, 95%, and 99% it is often impossible to obtain a high level of confidence unless the sample size is large.

Output - Example 1.1:

```
                        The UNIVARIATE Procedure
                        Variable:  sodium

                              Moments

        N                    40    Sum Weights              40
        Mean              76.07    Sum Observations      3042.8
        Std Deviation 2.28778249   Variance          5.23394872
        Skewness      0.40173009   Kurtosis          -0.6099696
        Uncorrected SS 231669.92   Corrected SS         204.124
        Coeff Variation 3.00747008 Std Error Mean    0.36173017


                      Basic Statistical Measures

            Location                      Variability

        Mean      76.07000    Std Deviation          2.28778
        Median    75.35000    Variance               5.23395
        Mode      73.30000    Range                  8.90000
                              Interquartile Range    3.30000

    NOTE: The mode displayed is the smallest of 2 modes with a count of 3.


                      Tests for Location: MuO=75

           Test          -Statistic-    -----p Value------

           Student's t   t  2.958006    Pr > |t|    0.0052
(1)        Sign          M       6.5    Pr >= |M|   0.0533
           Signed Rank   S     176.5    Pr >= |S|   0.0118


                        Quantiles (Definition 5)

                            95% Confidence Limits  -------Order Statistics-------
        Quantile    Estimate    Distribution Free  LCL Rank  UCL Rank  Coverage

        100% Max      81.00
        99%           81.00      80.3       81.0      38        40      32.35
        95%           80.40      78.9       81.0      36        40      82.35
        90%           79.30      78.0       81.0      32        40      96.97
        75% Q3        77.55      76.6       78.9      25        36      95.77
(2)     50% Median    75.35      75.0       77.1      14        27      96.15
        25% Q1        74.25      73.3       75.1       5        16      95.77
        10%           73.30      72.1       74.2       1         9      96.97
        5%            72.85      72.1       73.3       1         5      82.35
        1%            72.10      72.1       72.9       1         3      32.35
        0% Min        72.10
```

2
Two-Sample Methods

2.1   Two-Sample Permutation Test

Many nonparametric tests are variations of two-sample permutation tests. Suppose $m + n$ experimental units are randomly assigned to one of two treatments with m going to treatment 1 and n going to treatment 2. If there is no difference between the treatments, then all possible $\binom{m+n}{m}$ random assignments (or permutations) of the observations among the two treatments are equally likely to occur. Let T be a summary statistic that measures the difference between the two treatments, for instance, the difference between the two means. The distribution of the possible values of T obtained by permuting the observations among the treatments is called the *permutation distribution*. If T_{obs} is the observed value of T from the original data, then the *p*-value associated with T_{obs} is the proportion of the permutation distribution for which T is as extreme as or more extreme than T_{obs}.

Statistics such as the difference of means, the difference of medians, and the ratio of means may be used in permutation tests, although it is common to use the difference of means or an equivalent statistic. There are several statistics that are equivalent to the difference of means in the sense that the permutation *p*-values of these statistics are the same. These include the sum of the observations of one of the groups, the mean of one of the groups, the pooled two-sample *t*-statistic for testing differences of means, and, for two-sided tests, the one-way analysis of variance *F*-statistic, which is the square of the pooled two-sample *t*-statistic.

Large Sample Approximation
The theory of sampling from finite populations (Cochran, 1963) is used to determine the permutation mean and variance of the statistic $S = sum\ of\ m\ observations$. Let the population mean and variance of the $m + n$ data values be denoted by μ and σ^2, respectively. Then $E(S) = m\mu$ and $Var(S) = mn\sigma^2/(m+n-1)$. The statistic $Z = \left(S - E(S)\right)/\sqrt{Var(S)}$ has an approximate standard normal distribution from which an approximate *p*-value may be obtained.

SAS Implementation
The **NPAR1WAY** procedure will perform a permutation test using the sum of the scores of one of the samples as the test statistic.

Example 2.1.1. The data in Table 2.1.1 give the test scores of seven employees of a company, four of whom were assigned at random to a new method of instruction, while the other three were assigned at random to the traditional method. Suppose the company wishes to determine if the new method improves scores.

Table 2.1.1. Test scores of seven employees for comparison of methods of instruction.	
New method	37, 49, 55, 57
Traditional method	23, 31, 46

The following code requests a permutation test for the difference between two means. The **scores=data** option requests that the raw data be used for the test. The **anova** option in the **PROC NPAR1WAY** statement requests the parametric analysis of variance (equivalent to a pooled two-sample *t*-test). The **exact** statement requests that the *p*-value be based on all possible permutations of the data.

```
data ta2_1_1;
input method $ score;
datalines;
new 37
new 49
new 55
new 57
trad 23
trad 31
trad 46
;

proc npar1way data=ta2_1_1 anova scores=data;   /* Requests analysis using raw data */
class method;                                    /* Identifies a classification variable */
exact scores=data;                               /* Requests exact p-value using raw data */
var score;                                       /* Specifies analysis variable */
run;
```

The output gives the result of the analysis of variance **(1)** equivalent to a two-sample pooled *t*-test, yielding a two-sided *p*-value of 0.0916 (one-sided *p*-value is 0.0946/2 = 0.046). The test statistic is $t = \sqrt{F \ value} = \sqrt{4.3444} = 2.08$. The test statistic value for the permutation test **(2)** is $S = 100 = $ sum of values from the traditional instruction sample. (Note: SAS uses the sum the scores of the observations in the smaller of the two samples. If both samples have the same number of observations, then **PROC NPAR1WAY** sums those scores for the sample that appears first in the input data set.) Next is the large sample approximation **(3)** to the permutation test ($Z = -1.6702$, one-sided *p*-value 0.0474). The permutation test is given under the heading "Exact Test" **(4)** with a one-sided *p*-value of 0.0571. Finally, in **(5)** is an equivalent test to the two-sided test given in **(3)**, which can be extended to more than two samples (see Section 3.1).

***Output - Example 2.1.1*:**

```
                          The NPAR1WAY Procedure                    1
```

(1)
```
                    Analysis of Variance for Variable score
                         Classified by Variable method
```

method	N	Mean
new	4	49.500000
trad	3	33.333333

Source	DF	Sum of Squares	Mean Square	F Value	Pr > F
Among	1	448.047619	448.047619	4.3444	0.0916
Within	5	515.666667	103.133333		

```
                    Data Scores for Variable score                 2
                       Classified by Variable method
```

method	N	Sum of Scores	Expected Under H0	Std Dev Under H0	Mean Score
new	4	198.0	170.285714	16.593581	49.500000
trad	3	100.0	127.714286	16.593581	33.333333

```
                      Data Scores Two-Sample Test
```

(2)
```
                    Statistic (S)              100.0000
```

(3)
```
                    Normal Approximation
                    Z                          -1.6702
                    One-Sided Pr <  Z           0.0474
                    Two-Sided Pr > |Z|          0.0949
```

(4)
```
                    Exact Test
                    One-Sided Pr <=  S          0.0571
                    Two-Sided Pr >= |S - Mean|  0.1143
```

```
                      Data Scores One-Way Analysis
```

(5)
```
                    Chi-Square          2.7895
                    DF                  1
                    Pr > Chi-Square     0.0949
```

2.2   Random Sampling the Permutations

Even with a relatively small number of observations, the number of permutations will be large. For instance, if there are 10 observations in each of two treatments, then there are 184,756 permutations. Thus, it may not be feasible to obtain the complete permutation distribution of a statistic T. In such cases, a random sample of the permutations may be used to approximate the p-value of an observed statistic T_{obs}. With N randomly selected permutations, an approximate p-value may be obtained by computing the fraction of the N permutations for which the value of the statistic T is as extreme as or more extreme than the observed statistic T_{obs}. A 99% confidence interval for the actual p-value is

$$\hat{p} \pm 2.576 \sqrt{\frac{\hat{p}(1-\hat{p})}{N}} \text{ , where } \hat{p} \text{ is the estimated } p\text{-value.}$$

SAS Implementation

The **n=*number of samples*** option or the **mc** option, in the **exact** statement, can be used to request the p-value of the permutation test based on a random sample of permutations instead of all possible permutations. The **mc** option requests 10,000 randomly selected permutations.

Example 2.2.1. Table 2.2.1 gives times, in minutes, to observe a 2-liter runoff of water for soil tilled using two different methods.

Table 2.2.1. Minutes to observe 2-liter runoff for two treatments.	
Treatment 1	59.1 60.3 58.1 61.3 65.1 55.0 63.4 67.8
Treatment 2	60.1 62.1 59.3 55.0 54.6 64.4 58.7 62.5

The following code uses the **n=1000** option to request a p-value based on 1000 randomly selected permutations.

```
data ta2_2_1;
input treat minutes @@;
datalines;
1 59.1 1 60.3 1 58.1 1 61.3 1 65.1 1 55 1 63.4 1 67.8
2 60.1 2 62.1 2 59.3 2 55 2 54.6 2 64.4 2 58.7 2 62.5
;

proc npar1way data=ta2_2_1 scores=data;
class treat;
exact scores=data/n=1000;   /* Requests p-value based on a random sample of 1000 */
                            /* permutations                                      */
var minutes;
run;
```

The output is similar to that of Section 2.1, except that the "Exact Test" section is now replaced by the section labeled "Monte Carlo Estimates for the Exact Test" **(1)**. The estimated p-value is 0.183. Because this estimate is based on a random sample of permutations, we would expect to obtain slightly different values each time we run the test. Also given in the output are 99% confidence limits for the exact p-value using all permutations **(2)**. In this case, we are 99% confident that we would obtain a p-value between 0.1515 and 0.2145 if all possible permutations were to be used to compute the p-value. [*Note*: If it is desired to obtain the same p-value estimate each time the program is executed, then the **seed=*positive integer*** option can be used. The **mc** option is automatically invoked when the **seed= *positive integer*** option is used.]

Output - Example 2.2.1:

```
                    The NPAR1WAY Procedure

              Data Scores for Variable minutes
                 Classified by Variable treat

                        Sum of      Expected      Std Dev       Mean
     treat     N        Scores      Under HO      Under HO      Score
     ─────────────────────────────────────────────────────────────────
       1       8        490.10      483.40        7.538170      61.26250
       2       8        476.70      483.40        7.538170      59.58750


                  Data Scores Two-Sample Test

           Statistic                    490.1000
           Z                              0.8888
           One-Sided Pr > Z               0.1871
           Two-Sided Pr > |Z|             0.3741


            Monte Carlo Estimates for the Exact Test

           One-Sided Pr >= S
(1)        Estimate                       0.1830
(2)        99% Lower Conf Limit           0.1515
           99% Upper Conf Limit           0.2145

           Two-Sided Pr >= |S - Mean|
           Estimate                       0.3840
           99% Lower Conf Limit           0.3444
           99% Upper Conf Limit           0.4236

           Number of Samples                1000
           Initial Seed                    38280


                  Data Scores One-Way Analysis

           Chi-Square                     0.7900
           DF                                  1
           Pr > Chi-Square                0.3741
```

The **MULTTEST** procedure will also perform permutation tests for the difference between means. The **perm** option in the **PROC MULTTEST** statement requests a *p*-value based on the permutation distribution. Only *p*-values based on random samples of the permutations are permitted. If the **n=***number of samples* option is omitted, then the default is **n=20,000** random samples. The **n=1000** option requests 1000 randomly sampled permutations. The **contrast** statement must be used to identify the form of the means comparison. The **test** statement requests comparisons based on the means. The following code produces a permutation *p*-value for the data of Table 2.2.1:

```
proc multtest perm n=1000;        /* Requests 1000 random permutations */
class treat;
contrast 'treat1 vs treat2' -1 1 ;  /* Specifies contrast coefficients */
test mean(minutes);               /* Requests comparisons based on means */
run;
```

The one-sided *p*-value **(3)** is 0.393/2 = 0.1965. The "Raw" *p*-value reported of 0.3926/2 = 0.1963 is that of the independent samples pooled *t*-test.

Output - Example 2.2.1 (continued):

The Multtest Procedure

Model Information

Test for continuous variables:	Mean t-test
Tails for continuous tests:	Two-tailed
Strata weights:	None
P-value adjustment:	Permutation
Center continuous variables?	No
Number of resamples:	1000
Seed:	38651

Contrast Coefficients

	treat	
Contrast	1	2
treat1 vs treat2	-1	1

Continuous Variable Tabulations

Variable	treat	NumObs	Mean	Standard Deviation
minutes	1	8	61.2625	4.0858
minutes	2	8	59.5875	3.4848

p-Values

	Variable	Contrast	Raw	Permutation
(3)	minutes	treat1 vs treat2	0.3926	0.3930

The **MULTTEST** procedure will also compute *p*-values based on bootstrap sampling (see Section 2.6) and will perform multiple pairwise comparisons for several samples, using permutation and bootstrap sampling (see Section 3.3).

2.3   Wilcoxon Rank-Sum Test

If the $m + n$ observations in the combined data set are distinct, then the ranks are assigned as 1 for the smallest, 2 for the next smallest, and so forth. If there are ties among the observations, the ranks of the tied observations are averaged. For instance, if the observations are 10, 12, 13, 13, and 15, then the ranks are 1, 2, 3.5, 3.5, and 5. The Wilcoxon rank-sum test is just a permutation test applied to the ranks where the statistic is the sum of ranks for one of the treatments. The large sample approximation is the same as the approximation for permutation tests in Section 2.1 except that the formulas for $E(S)$ and $Var(S)$ are applied to ranks instead of to the original observations.

SAS Implementation

The Wilcoxon rank-sum test can be requested by specifying the **wilcoxon** option in the **PROC NPAR1WAY** statement. If an exact *p*-value is desired, then the **wilcoxon** option is specified in the **exact** statement as well (if an **exact** statement is used, the **wilcoxon** option does not also need to be specified in the **PROC NPAR1WAY** statement). Average ranks are assigned to tied observations. The **n=***number of samples* and **mc** options can also be used for random sampling the permutations (see Section 2.2.)

Example 2.3.1. Table 2.3.1 gives dry weights, in kg, of strawberry plants, where seven were treated with an herbicide and nine were untreated.

Table 2.3.1. Dry weights (in kg) of strawberry plants.	
Untreated	0.55 0.67 0.63 0.79 0.81 0.85 0.68
Treated	0.65 0.59 0.44 0.60 0.47 0.58 0.66 0.52 0.51

The following code requests an exact *p*-value for the Wilcoxon rank-sum test.

```
data ta2_3_1;
input treat $ weight @@;
datalines;
u .55 u .67 u .63 u .79 u .81 u .85 u .68
t .65 t .59 t .44 t .6 t .47 t .58 t .66 t .52 t .51
;

proc npar1way data=ta2_3_1 wilcoxon;    /* Requests analysis on ranks */
class treat;
exact wilcoxon;                         /* Requests exact p-value for Wilcoxon test */
var weight;
run;
```

The output is similar to that of the permutation test for the difference between two means, because SAS performs the same procedure, only on the ranks of the data. The test statistic, *S*, is the sum of the ranks for the smaller sample **(1)**. The first test given **(2)** is that based on the large sample approximation to the permutation distribution. The one-sided *p*-value is 0.0048. The "Exact Test" section **(3)** gives the exact one-sided *p*-value of 0.0039 for the Wilcoxon rank-sum test.

Output - Example 2.3.1:

The NPAR1WAY Procedure

Wilcoxon Scores (Rank Sums) for Variable weight
Classified by Variable treat

treat	N	Sum of Scores	Expected Under H0	Std Dev Under H0	Mean Score
u	7	84.0	59.50	9.447222	12.000000
t	9	52.0	76.50	9.447222	5.777778

Wilcoxon Two-Sample Test

(1)
Statistic (S)   84.0000

(2)
Normal Approximation
Z   2.5934
One-Sided Pr > Z   0.0048
Two-Sided Pr > |Z| 0.0095

t Approximation
One-Sided Pr > Z   0.0102
Two-Sided Pr > |Z| 0.0204

(3)
Exact Test
One-Sided Pr >= S   0.0039
Two-Sided Pr >= |S - Mean| 0.0079

Kruskal-Wallis Test

Chi-Square   6.7255
DF   1
Pr > Chi-Square   0.0095

Wilcoxon Rank-Sum Test vs. Pooled Two-Sample *t*-Test

Suppose that $F_1(x)$ and $F_2(x)$ are the distribution functions of two populations and we test $H_0 : F_1(x) = F_2(x)$ against the shift alternative $H_a : F_1(x) = F_2(x-d)$. The level of significance of the Wilcoxon rank-sum test is exact in the sense that does not depend on the functional form of F_1 and F_2. On the other hand, the nominal level of significance of the pooled two-sample *t*-test can only be assumed to be correct when the populations have a normal distribution. Fortunately, the nominal level of significance of the *t*-test is often quite close to the correct level of significance even if the distributions are non-normal, provided the distributions have finite variance. For large samples we may appeal to the Central Limit Theorem, while for small samples numerous simulation studies have shown the *t*-test to be robust with respect to Type I errors. Thus, the choice between the Wilcoxon rank-sum test and the pooled two-sample *t*-test often comes down to the issue of which one is more powerful.

If observations are from normal distributions, or if the distributions have light tails, such as the uniform distribution, then the two-sample *t*-test will be more powerful than the Wilcoxon rank-sum test. On the other hand, if the distributions have heavy tails, such as the Laplace distribution or the Cauchy distribution, then the Wilcoxon rank-sum test will be more powerful (Blair and Higgins, 1980). There is the intuitive notion that ranking must necessarily lead to a less powerful test because information is "thrown away" when observations are replaced with their ranks, but this is not the case. The Wilcoxon test gives less weight to extreme observations, and when such observations are likely to occur, the Wilcoxon test will be more powerful than the *t*-test. Under rather general conditions, the permutation test has the same asymptotic power as the *t*-test (Pitman, 1937) and thus it will generally do well or poorly relative to the rank-sum test in the same circumstances as the *t*-test. The following example illustrates this point.

Example 2.3.2. The data in Table 2.3.2 are the numbers of brothers and sisters of students in an introductory statistics class. The data were classified into two groups according to the size of the students' home community. A look at the data shows that rural students tend to have more brothers and sisters than urban students except that one urban student has 8 brothers and sisters. There is no justification for the deletion of this value from the data set because it is not an error. However, it greatly affects the analysis. The *p*-values for the one-way ANOVA **(1)** and the permutation *t*-test **(3)** are comparable, being 0.1094 and 0.1131, respectively, whereas the *p*-value for the Wilcoxon rank-sum test **(2)** is 0.0010. In this case, the Wilcoxon test can detect an apparent difference between two groups, but the other two tests cannot, even at the 10% level of significance. If the extreme observation 8 is deleted, then the *p*-values are 0.0013, 0.0006, and 0.0001 for the *t*-test, permutation test, and Wilcoxon test, respectively.

Table 2.3.2. Number of siblings of students in rural and urban areas.	
Rural (1)	3 2 1 1 2 1 3 2 2 2 2 5 1 4 1 1 1 1 6 2 2 2 1 1
Urban (2)	1 0 1 1 0 0 1 1 1 8 1 1 1 0 1 1 2

Output - Example 2.3.2:

```
                      The NPAR1WAY Procedure
              Analysis of Variance for Variable Siblings
                     Classified by Variable Group

            Group              N                Mean

              1                24             2.041667
              2                17             1.235294


     Source    DF   Sum of Squares   Mean Square    F Value    Pr > F

     Among      1         6.470648      6.470648     2.6841     0.1094
     Within    39        94.017157      2.410696


              Average scores were used for ties.

       Wilcoxon Scores (Rank Sums) for Variable Siblings
                  Classified by Variable Group

                        Sum of      Expected      Std Dev        Mean
     Group      N       Scores      Under H0      Under H0       Score

       1        24      614.50        504.0       34.850342    25.604167
       2        17      246.50        357.0       34.850342    14.500000

           Average scores were used for ties.
              Wilcoxon Two-Sample Test
         Statistic (S)                   246.5000
         Exact Test
         One-Sided Pr <=  S              4.144E-04
         Two-Sided Pr >= |S - Mean|      0.0010
```

(1) *(at Among/Within rows)*

(2) *(at Two-Sided Pr row)*

Output – Example 2.3.2 (continued):

```
              Data Scores for Variable Siblings
                   Classified by Variable Group

                      Sum of     Expected      Std Dev        Mean
    Group      N      Scores     Under H0      Under H0       Score

      1       24       49.0      40.975610     4.999941     2.041667
      2       17       21.0      29.024390     4.999941     1.235294

              Data Scores Two-Sample Test

         Statistic (S)                    21.0000

         Exact Test
         One-Sided Pr <=  S                0.0637
         Two-Sided Pr >= |S - Mean|        0.1131
```

(3)

2.4 Other Scoring Systems

Ranks can be thought of as scores that are used in place of observations. To motivate other scoring systems, consider a particular way to think about ranks. Suppose we take a random sample of size n from the uniform distribution on the interval $[0, n+1]$. Let $U_{(1)} < U_{(2)} < \cdots < U_{(n)}$ denote the order statistics of this random sample; that is, $U_{(1)}$ is the smallest observation, $U_{(2)}$ is the next smallest, and so on. It can be shown that the ranks are the expected values of the order statistics, that is $E(U_{(i)}) = i$. Other scoring systems are based on expected order statistics of distributions other than the uniform.

The *exponential* or *Savage scores* are obtained by replacing the ith smallest observation with $E(W_{(i)}) - 1$ where $W_{(i)}$ is the ith ordered observation from a standard exponential distribution. *Normal scores* are the expected values of the ordered observations from a standard normal distribution. These scores may be approximated by *Van der Waerden scores*. The ith Van der Waerden score is defined as $V_{(1)} = \Phi^{-1}\left(\dfrac{i}{n+1}\right)$, where Φ is the standard normal cumulative distribution function. If there are ties in the data, then the scores are averaged for the tied observations. Nonparametric tests are obtained by doing permutation tests on the scores. The purpose of using scores other than ranks is to produce tests that are more powerful than the Wilcoxon rank-sum test in certain circumstances. See Lehmann (1975) for a discussion of the issues.

SAS Implementation

The **NPAR1WAY** procedure allows analysis using Van der Waerden (quantile normal) scores and Savage (exponential) scores using the **vw** and **savage** options, respectively, in the **PROC NPAR1WAY** statement. Exact p-values based on tests using these scores can be obtained by using the same options with the **exact** statement.

Example 2.4.1. Table 2.4.1 gives the amounts of cerium, in parts per million, in samples of granite and basalt.

Table 2.4.1. Amounts of cerium (in ppm) in samples of granite and basalt.	
Granite	33.63 39.86 69.32 42.13 58.36 74.11
Basalt	26.15 18.56 17.55 9.84 28.29 34.15

The following code can be used to compute exact *p*-values using Van der Waerden and Savage scores.

```
data ta2_4_1;
input treat $ cerium @@;
datalines;
g 33.63 g 39.86 g 69.32 g 42.13 g 58.36 g 74.11
b 26.15 b 18.56 b 17.55 b 9.84 b 28.29 b 34.15
;

proc npar1way data=ta2_4_1 vw savage;   /* Requests analysis on VW and Savage scores */
class treat;
exact vw savage;                        /* Requests exact p-values              */
var cerium;
run;
```

The output is identical to that of Example 2.1.1, except that the permutation test is performed using first the Van der Waerden, then the Savage scores, instead of the raw data.

Output – Example 2.4.1:

```
                        The NPAR1WAY Procedure

            Van der Waerden Scores (Normal) for Variable cerium
                       Classified by Variable treat

                      Sum of     Expected      Std Dev        Mean
     treat     N      Scores     Under HO     Under HO       Score

     x         6    3.881694          0.0     1.470476    0.646949
     y         6   -3.881694          0.0     1.470476   -0.646949


                   Van der Waerden Two-Sample Test

                   Statistic (S)              3.8817

                   Normal Approximation
                   Z                          2.6398
                   One-Sided Pr >  Z          0.0041
                   Two-Sided Pr > |Z|         0.0083

                   Exact Test
                   One-Sided Pr >=  S         0.0022
                   Two-Sided Pr >= |S - Mean| 0.0043
```

Output – Example 2.4.1 (continued):

```
                  Van der Waerden One-Way Analysis

                    Chi-Square              6.9683
                    DF                           1
                    Pr > Chi-Square         0.0083
```

```
                    The NPAR1WAY Procedure

          Savage Scores (Exponential) for Variable cerium
                    Classified by Variable treat
```

treat	N	Sum of Scores	Expected Under H0	Std Dev Under H0	Mean Score
x	6	3.752597	0.0	1.557690	0.625433
y	6	-3.752597	0.0	1.557690	-0.625433

```
                  Savage Two-Sample Test

        Statistic (S)               3.7526

        Normal Approximation
        Z                           2.4091
        One-Sided Pr >  Z           0.0080
        Two-Sided Pr > |Z|          0.0160

        Exact Test
        One-Sided Pr >=  S          0.0022
        Two-Sided Pr >= |S - Mean|  0.0043
```

```
                  Savage One-Way Analysis

                    Chi-Square              5.8037
                    DF                           1
                    Pr > Chi-Square         0.0160
```

2.5 Tests for Equality of Scale Parameters and Omnibus Tests

Various scoring systems may be used to detect differences in scale parameters of two distributions. Let $F_1(x)$ and $F_2(x)$ denote the distribution functions of the two populations, and assume that the two distributions are continuous and have the same known median m. We test the null hypothesis $F_1(x) = F_2(x)$ against the alternative $F_1(x-m) = F_2(c(x-m))$, $c > 0$. Intuitively, the two distributions are identical except for a possible difference in scale parameter.

The Siegel-Tukey scoring system assigns 1 to the smallest observation, 2 to the largest observation, 3 to the next largest, 4 to the second smallest, and so on. The Ansari-Bradley scoring system assigns 1 to the smallest and largest observations, 2 to the next smallest and next largest observations, and so on. The Klotz scores are the squares of the Van der Waerden scores, and the Mood scores are squares of the deviations of the ranks from the mean of the ranks, that is,

$\left(i - \dfrac{n+1}{2} \right)^2$. Average scores are assigned to ties. Tests are carried out as permutation tests on the scores. Omnibus tests look for overall differences between two distribution functions. These tests are based on the empirical distribution function (EDF). If $X_1, X_2, \ldots, X_n$ are a random sample of size n from a population with distribution function $F(x)$, the EDF is

$\hat{F}(x) = $ *fraction of X_i's* $\leq x$. Let the EDFs of the two treatments be denoted $\hat{F}_i(x)$, $i = 1, 2$. The Kolmogorov-Smirnov (KS) test is based on $\max_x \left| \hat{F}_1(x) - \hat{F}_2(x) \right|$, the maximum difference between the two EDFs taken over the two samples. Kuiper's test is based on

$\max_x \left(F_1(x) - F_2(x) \right) - \min_x \left(F_1(x) - F_2(x) \right)$, where the maximum and minimum are taken over the two samples. The Cramer-von Mises statistic CM is proportional to the squared distance between the EDFs:

$$CM = c \sum_{i=1}^{m} \left(\hat{F}_1(x_i) - \hat{F}_2(x_i) \right)^2 + c \sum_{j=1}^{n} \left(\hat{F}_1(y_j) - \hat{F}_2(y_j) \right)^2,$$

where $c = \dfrac{mn}{(m+n)^2}$ and the sums are taken over the data in the respective samples.

SAS Implementation

The **NPAR1WAY** procedure can perform tests based on Siegel-Tukey, Ansari-Bradley, Klotz and Mood scores for testing equality of scale parameters. In addition, three statistics based on the empirical distribution function—the Kolmogorov-Smirnov, Cramer-von Mises, and Kuiper—are available. The following options are specified in the **PROC NPAR1WAY** statement and in the **exact** statement if exact *p*-values are desired:

Option	Test performed
st	Siegel-Tukey
ab	Ansari-Bradley
klotz	Klotz
mood	Mood
edf	Kolmogorov-Smirnov, Cramer-von Mises, Kuiper
ks (available in versions 8.2 and higher)	Kolmogorov-Smirnov

Notes: No test is performed for the Cramer-von Mises statistic—only the CM statistic and its associated asymptotic value are given. When the **edf** option is used in the **exact** statement, an exact *p*-value is computed only for the Kolmogorov-Smirnov test. Alternatively, the **ks** option can be used to request only the Kolmogorov-Smirnov test.

Example 2.5.1. Table 2.5.1 gives ounces of beverages in containers measured before and after a process repair.

Table 2.5.1. Ounces of beverage in containers.

Treatment 1 (before)	16.55 15.36 15.94 16.43 16.01
Treatment 2 (after)	16.05 15.98 16.10 15.88 15.91

The following code requests all of the tests mentioned above for the data of Table 2.5.1, and it produces the output below. The output is similar in format to that of the previous examples in this chapter.

data ta2_5_1;
input treatment ounces @@;
datalines;
1 16.55 1 15.36 1 15.94 1 16.43 1 16.01
2 16.05 2 15.98 2 16.10 2 15.88 2 15.91
;

proc npar1way data=ta2_5_1 st ab klotz mood edf;   /* *Requests the tests* */
class treatment;
exact st ab klotz mood edf;   /* *Requests exact p-values* */
var ounces;
run;

Output – Example 2.5.1:

The NPAR1WAY Procedure

Siegel-Tukey Scores for Variable ounces
Classified by Variable treatment

treatment	N	Sum of Scores	Expected Under H0	Std Dev Under H0	Mean Score
1	5	24.0	27.50	4.787136	4.80
2	5	31.0	27.50	4.787136	6.20

Siegel-Tukey Two-Sample Test

Statistic (S)   24.0000

Normal Approximation
Z   -0.6267
One-Sided Pr < Z   0.2654
Two-Sided Pr > |Z| 0.5309

Exact Test
One-Sided Pr <= S   0.2738
Two-Sided Pr >= |S - Mean| 0.5476

Z includes a continuity correction of 0.5.

Siegel-Tukey One-Way Analysis

Chi-Square   0.5345
DF   1
Pr > Chi-Square   0.4647

Output – Example 2.5.1 (continued)**:**

```
           Ansari-Bradley Scores for Variable ounces
              Classified by Variable treatment

                        Sum of    Expected    Std Dev      Mean
 treatment       N      Scores    Under HO    Under HO     Score

 1               5       13.0        15.0     2.357023      2.60
 2               5       17.0        15.0     2.357023      3.40


              Ansari-Bradley Two-Sample Test

       Statistic (S)            13.0000

       Normal Approximation
       Z                        -0.8485
       One-Sided Pr <  Z         0.1981
       Two-Sided Pr > |Z|        0.3961

       Exact Test
       One-Sided Pr <=  S        0.2698
       Two-Sided Pr >= |S - Mean|  0.5397


              Ansari-Bradley One-Way Analysis

          Chi-Square        0.7200
          DF                     1
          Pr > Chi-Square   0.3961


              Klotz Scores for Variable ounces
              Classified by Variable treatment

                        Sum of    Expected    Std Dev      Mean
 treatment       N      Scores    Under HO    Under HO     Score

 1               5     4.525364   3.108188    1.073814   0.905073
 2               5     1.691011   3.108188    1.073814   0.338202


                 Klotz Two-Sample Test

       Statistic (S)             4.5254

       Normal Approximation
       Z                         1.3198
       One-Sided Pr >  Z         0.0935
       Two-Sided Pr > |Z|        0.1869

       Exact Test
       One-Sided Pr >=  S        0.1349
       Two-Sided Pr >= |S - Mean|  0.2698


                 Klotz One-Way Analysis

          Chi-Square        1.7418
          DF                     1
          Pr > Chi-Square   0.1869
```

Output – Example 2.5.1 (continued):

Mood Scores for Variable ounces
Classified by Variable treatment

treatment	N	Sum of Scores	Expected Under H0	Std Dev Under H0	Mean Score
1	5	55.250	41.250	12.110601	11.050
2	5	27.250	41.250	12.110601	5.450

Mood Two-Sample Test

Statistic (S)	55.2500

Normal Approximation
Z	1.1560
One-Sided Pr > Z	0.1238
Two-Sided Pr > |Z|	0.2477

Exact Test
One-Sided Pr >= S	0.1508
Two-Sided Pr >= |S - Mean|	0.3016

Mood One-Way Analysis

Chi-Square	1.3364
DF	1
Pr > Chi-Square	0.2477

Kolmogorov-Smirnov Test for Variable ounces
Classified by Variable treatment

treatment	N	EDF at Maximum	Deviation from Mean at Maximum
1	5	0.60	-0.447214
2	5	1.00	0.447214
Total	10	0.80	

Maximum Deviation Occurred at Observation 8
Value of ounces at Maximum = 16.10

Kolmogorov-Smirnov Two-Sample Test

KS	0.2000
D- = max (F2 - F1)	0.4000
D = max |F1 - F2|	0.4000

Asymptotic Test
KSa	0.6325
One-Sided Pr > D-	0.4493
Two-Sided Pr > D	0.8186

Exact Test
One-Sided Pr >= D-	0.4762
Two-Sided Pr >= D	0.8730

Output – Example 2.5.1 (continued):

```
         Cramer-von Mises Test for Variable ounces
              Classified by Variable treatment


                                        Summed Deviation
       treatment             N             from Mean

       1                     5                0.0450
       2                     5                0.0450


         Cramer-von Mises Statistics (Asymptotic)
            CM  0.009000     CMa  0.090000




             Kuiper Test for Variable ounces
             Classified by Variable treatment


                                         Deviation
         treatment             N         from Mean

         1                     5            0.20
         2                     5            0.40


         Kuiper Two-Sample Test (Asymptotic)
    K  0.600000     Ka  0.948683     Pr > Ka  0.8796
```

2.6 Nonparametric Bootstrap Methods for Two Sample Inference

Bootstrap resampling is a process of sampling from the data in a way that simulates sampling from the original population. Bootstrap resampling is nonparametric if the sampling is done from the data or the empirical distribution function; it is parametric if a functional form is imposed on the distribution from which the sample is selected.

2.6.1 Two-sample test using bootstrap resampling

Let the observed data from the two samples be denoted $x_1, x_2,..., x_m$ and $y_1, y_2,..., y_n$. Compute the observed t-statistic and p-value from the data. Let e_{ix} and e_{iy} denote the residuals defined by

$e_{ix} = x_i - \bar{x}$ and $e_{jy} = y_j - \bar{y}$. Pool the residuals together and select $m+n$ residuals with replacement. This is a bootstrap resampling of residuals. Assign m to treatment 1 and n to treatment 2, and compute the pooled two-sample t-statistic and p-value for the bootstrap sample. Repeat this procedure N times. The fraction of the bootstrap t-statistics as extreme as or more extreme than the observed t-statistic is the bootstrap p-value. Equivalently, the p-value is the fraction of the bootstrap p-values that are less than or equal to the observed p-value. This procedure is implemented in **PROC MULTTEST**.

The **MULTTEST** procedure allows for random sampling the permutations so that a two-sample permutation test may also be done with this procedure (see Section 2.2). Both the bootstrap and permutation procedures involve resampling the data. Bootstrap resampling is done on the residuals with replacement, and permutation resampling is done on the original data without replacement. In other words, random sampling the permutations is equivalent to randomly selecting *m* of the combined observations without replacement for treatment 1 and assigning the remaining *n* to treatment 2.

SAS Implementation

The following code illustrates the use of the **MULTTEST** procedure to compute a *p*-value for the difference between means using bootstrap resampling for the data of Table 2.2.1. The syntax is identical to that of the permutation test of Section 2.2, except that the option **boot** is used in the **PROC MULTTEST** statement instead of **perm**. Also, the number of randomly selected bootstrap samples is not specified, thus the default of **n=20,000** is used.

```
data ta2_2_1;
input treat minutes @@;
datalines;
1 59.1 1 60.3 1 58.1 1 61.3 1 65.1 1 55 1 63.4 1 67.8
2 60.1 2 62.1 2 59.3 2 55 2 54.6 2 64.4 2 58.7 2 62.5
;

proc multtest data= ta2_2_1 boot;  /* "boot" option requests bootstrap p-values */
class treat;
contrast 'treat1 vs treat2' -1 1;
test mean(minutes);
run;
quit;
```

The one-sided *p*-value based on 20,000 bootstrap samples is 0.3914/2 = 0.1957.

Output – Section 2.6 Example:

```
                    The Multtest Procedure

                    Model Information

        Test for continuous variables:        Mean t-test
        Tails for continuous tests:           Two-tailed
        Strata weights:                       None
        P-value adjustment:                   Bootstrap
        Center continuous variables?          Yes
        Number of resamples:                  20000
        Seed:                                 45433

                    Contrast Coefficients
                                    treat

        Contrast                   1              2

        treat1 vs treat2          -1              1
```

Output – Section 2.6 Example (continued):

                    Continuous Variable Tabulations

| | | | | Standard |
Variable	treat	NumObs	Mean	Deviation
minutes	1	8	61.2625	4.0858
minutes	2	8	59.5875	3.4848

                            p-Values

Variable	Contrast	Raw	Bootstrap
minutes	treat1 vs treat2	0.3926	0.3914

2.6.2 Other bootstrap procedures in SAS

For those familiar with the SAS Macro Language, the **%BOOTCI** macro can be used to compute nonparametric bootstrap confidence intervals.

3
K-Sample Methods

3.1 *K*-Sample Permutation Tests

Suppose an experiment has k treatments or groups, and n_i observations are taken from treatment i, and let Y_{ij} denote the jth observation from the ith treatment, $i = 1, 2, ..., k$, $j = 1, 2, ..., n_i$. Let $\overline{Y}_i$ and S_i^2 denote the mean and sample variance of the observations from the ith treatment. Let $\overline{Y}$ denote the average of all the observations, and let $N = \sum_{i=1}^{k} n_i$. The total sum of squares

is $SS_{total} = \sum_{i=1}^{k} \sum_{j=1}^{n_i} (Y_{ij} - \overline{Y})^2$. The sum of squares for the treatments is $SST = \sum_{i=1}^{k} n_i (\overline{Y}_i - \overline{Y})^2$, and the

sum of squares errors is $SSE = \sum_{i=1}^{k} (n_i - 1) S_i^2$. When observations come from normal distributions

with equal population variances, the test statistic for the null hypothesis of equal population means is $F = \dfrac{SST/(k-1)}{SSE/(N-k)}$, where the reference distribution is the F-distribution with $k-1$ degrees of freedom for the numerator and $N - k$ degrees of freedom for the denominator.

In cases in which the observations do not come from normal distributions, we may apply a permutation test. Let $F_i(x)$ denote the distribution function of observations from treatment i, $i = 1, 2, ..., k$. The null hypothesis is $H_0: F_1(x) = F_2(x) = ... = F_k(x)$, that is, equality of distributions. The alternative is $H_a: F_i(x) \leq F_j(x)$ or $F_i(x) \geq F_j(x)$ for at least one pair (i, j) with strict inequality holding for at least one x. Intuitively, the alternative states that there are at least two treatments for which observations from one tend to be larger than observations from the other. In carrying out the permutation F-test, we permute the observations among the treatments, and for each permutation we obtain the F-statistic or an equivalent statistic. The permutation p-value is the fraction of F's greater than or equal to F_{obs} where F_{obs} is the observed F-statistic.

SAS Implementation

The **NPAR1WAY** procedure will perform a permutation test using the statistic $(N-1) SST / SS_{total}$, which is equivalent to using the F-statistic.

Example 3.1.1. The data in Table 3.1.1 give samples of size $n = 5$ from normally distributed populations with standard deviation $\sigma = 9$ and respective means $\mu_1 = 15$, $\mu_2 = 25$, and $\mu_3 = 30$.

> **Table 3.1.1.** Samples from normally distributed populations with standard deviation $\sigma = 9$ and respective means $\mu_1 = 15$, $\mu_2 = 25$, and $\mu_3 = 30$.
>
> | | |
> |---|---|
> | Sample 1: | 6.08, 22.29, 7.51, 34.36, 23.68 |
> | Sample 2: | 30.45, 22.71, 44.52, 31.47, 36.81 |
> | Sample 3: | 32.04, 28.03, 32.74, 23.84, 29.64 |

The following code generates the standard analysis of variance, as well as a permutation F-test. The **scores=data** option requests analysis on the raw data values, and the **exact** statement requests the p-value be based on all possible permutations of the raw data.

```
data ta3_1_1;
input treat resp @@;
cards;
1 6.08 1 22.29 1 7.51 1 34.36 1 23.68
2 30.45 2 22.71 2 44.52 2 31.47 2 36.81
3 32.04 3 28.03 3 32.74 3 23.84 3 29.64
;

proc npar1way data=ta3_1_1
      anova              /* Requests analysis of variance */
      scores=data;       /* Requests analysis on raw data */
class treat;
exact scores=data;       /* Requests exact test on raw data */
var resp;
run;
```

The parametric test is given in **(1)**. The p-value is 0.0533. The permutation F-test is given in **(2)**. The exact p-value 0.0513. [Note: Computation of the exact p-value may take a considerable amount of time, because even for this relatively small data set there are more than 700,000 permutations. If this occurs, it may be desirable to use the **mc** option (see Section 2.2).]

Output – Example 3.1.1:

The NPAR1WAY Procedure

(1)   Analysis of Variance for Variable resp
                  Classified by Variable treat

treat	N	Mean
1	5	18.7840
2	5	33.1920
3	5	29.2580

Source	DF	Sum of Squares	Mean Square	F Value	Pr > F
Among	2	554.619160	277.309580	3.7814	0.0533
Within	12	880.011480	73.334290		

Output – Example 3.1.1 (continued):

(2)

```
                        Data Scores for Variable resp
                        Classified by Variable treat

                   Sum of      Expected      Std Dev        Mean
    treat    N     Scores      Under HO      Under HO       Score

    1        5     93.920      135.390       18.481848      18.7840
    2        5     165.960     135.390       18.481848      33.1920
    3        5     146.290     135.390       18.481848      29.2580


                   Data Scores One-Way Analysis

    Chi-Square                          5.4123
    DF                                       2
    Asymptotic Pr >  Chi-Square         0.0668
    Exact       Pr >= Chi-Square        0.0513
```

3.2   The Kruskal-Wallis Test

The Kruskal-Wallis test is equivalent to a permutation F-test applied to ranks. The scores for the Kruskal-Wallis test are the ranks of the data with average ranks being used for tied observations. In the case of two treatments, it is equivalent to a two-sided Wilcoxon rank-sum test. The usual form of the test statistic is $KW = (N-1)SST/SS_{total}$, where the sums of squares formulas are applied to the ranks instead of the original observations. This form of the test statistic has a limiting chi-square distribution with $k-1$ degrees of freedom. It is also the form used for other scores such as the Savage or Van der Waerden scores. The power properties of the Kruskal-Wallis test are similar to the power properties of the Wilcoxon rank-sum test. In particular, when distributions have heavy tails, the Kruskal-Wallis test will be more powerful than the normal-theory F-test, and it is not adversely affected by outliers in the way the F-test is.

SAS Implementation

The **scores=wilcoxon** option in the **PROC NPAR1WAY** statement requests analysis on ranks and thus the Kruskal-Wallis test. Using the **scores=wilcoxon** option in the **exact** statement requests the exact p-value based on all (or a random sample of) permutations of the data.

Example 3.2.1. Three preservatives and a control were compared in terms of their ability to inhibit the growth of bacteria. Samples were treated with one of three preservatives or left untreated for the control, and bacteria counts were made 48 hours later. The data in Table 3.2.1 are the logarithms of the counts.

Table 3.2.1. Logarithms of bacteria counts for three preservatives and a control.						
Control:	4.302	4.017	4.049	4.176		
Preservative 1:	2.021	3.190	3.250	3.276	3.292	3.267
Preservative 2:	3.397	3.552	3.630	3.578	3.612	
Preservative 3:	2.699	2.929	2.785	2.176	2.845	2.913

The following code requests the Kruskal-Wallis test on the data of Table 3.2.1.

```
data ta3_2_1;
input treat $ logcount @@;
cards;
C 4.302 C 4.017 C 4.049 C 4.176
P1 2.021 P1 3.190 P1 3.250 P1 3.276 P1 3.292 P1 3.267
P2 3.973 P2 3.552 P2 3.630 P2 3.578 P2 3.612
P3 2.699 P3 2.929 P3 2.785 P3 2.176 P3 2.845 P3 2.913
;

proc npar1way data=ta3_2_1
      wilcoxon;                    /* Requests analysis on ranks */
class treat;
      exact wilcoxon /             /* Requests exact test on ranks */
      mc;                          /* Requests 10,000 random permutations */
var logcount;
run;
quit;
```

The *p*-value based on the chi-square approximation is 0.0007 as given in **(1)**. The *p*-value based on 10,000 randomly selected permutations is less than 0.0001, with an upper 99% confidence limit of 0.00046 as given in **(2)**.

Output – Example 3.2.1:

```
                         The NPAR1WAY Procedure

              Wilcoxon Scores (Rank Sums) for Variable logcount
                        Classified by Variable treat

                          Sum of      Expected      Std Dev        Mean
        treat      N      Scores      Under H0      Under H0       Score
        ---------------------------------------------------------------------
        C          4       78.0        44.0        11.165423      19.50
        P1         6       51.0        66.0        12.845233       8.50
        P2         5       75.0        55.0        12.110601      15.00
        P3         6       27.0        66.0        12.845233       4.50
```

Output – Example 3.2.1 (continued):

(1)
```
                    Kruskal-Wallis Test

           Chi-Square         17.1429
           DF                       3
           Pr > Chi-Square    0.0007
```

(2)
```
           Monte Carlo Estimate for the Exact Test

           Pr >= Chi-Square
           Estimate                0.0000
           99% Lower Conf Limit    0.0000
           99% Upper Conf Limit    4.604E-04

           Number of Samples         10000
           Initial Seed              83057
```

3.3   Multiple Comparisons

If there are differences among treatments, we would like to know where differences exist. Various procedures are available for comparing treatments based on which error rates one wishes to control. See Westfall et al. (1999) for a discussion of multiple comparisons in SAS. We will consider only the experiment-wise error rate, which is the probability of declaring that there is a difference among two or more treatments when no differences exist among any of the treatments.

There are two simple nonparametric procedures that may be carried out using **PROC NPAR1WAY**. If an overall test such as the Kruskal-Wallis test shows that there are differences among the treatments at level of significance α, then perform all pairwise tests at level of significance α using the same scores as the overall test. Because no tests are performed unless the overall test is significant, the experiment-wise error rate is no more than α. The procedure does not work well in controlling error rates under a partial null hypothesis. For instance, suppose that there is a difference between treatments 1 and 2 that causes the overall test to be significant but there are no differences among the other treatments. The significance of the overall test gives a license to look for differences among other treatments, and if there are a large number of treatments, significant differences are likely to occur just by chance.

A second and more conservative method for doing pairwise comparisons is the Bonferroni method. If there are k treatments, then there are $k(k-1)/2$ pairwise comparisons. If we wish to have an experiment-wise error rate of α, then we do each pairwise comparison at level of significance $\alpha/[k(k-1)/2]$. With this procedure, under any partial null hypothesis, the error rate is no more than α. If p_{ij} is the p-value for the pairwise comparison of treatments i and j, then the Bonferroni-adjusted p-value is obtained by multiplying p_{ij} by $k(k-1)/2$.

Another multiple comparison procedure, a version of which is implemented in **PROC MULTTEST**, is the minimum p-value method. Let T_{ij} denote a test statistic for comparing treatments i and j, $i = 1, 2, ..., k$, $j \neq i$. Imagine repeating the experiment over and over and for each repetition and obtaining p-values for the T_{ij}'s. In this sense, the p-values may be regarded as random variables. Let G_{min} denote the distribution function of the minimum of these p-values under the hypothesis that there are no differences among the treatments. For an observed p-value p_{ij} for treatments i and j, the adjusted p-value is $G_{min}(p_{ij})$. It is the probability that the minimum *p*-value, considered as a random variable, is less than or equal to the observed *p*-value under the null hypothesis of no differences among treatments. Treatments *i* and *j* would be declared to be significantly different at level of significance α if $G_{min}\left(p_{ij}\right) \le \alpha$. **PROC MULTTEST** will compute the permutation distribution of the minimum of the *p*-values so that the reference distribution $G_{min}\left(x\right)$ is a permutation distribution.

Suppose the T_{ij}'s are *t*-statistics with the same degrees of freedom and that the *t*-tests are two-sided. Let $T_{max\ abs}$ denote the maximum of the absolute values of the *t*-statistics. An adjusted *p*-value for an observed *t*-statistic t_{ij} is $P\left(T_{max\ abs} \ge |t_{ij}|\right)$. Because the *t*-statistics have the same degrees of freedom, there is a one-to-one correspondence between the *t*-statistics and their respective *p*-values. An adjusted *p*-value obtained this way is equivalent to the adjusted *p*-value obtained by the minimum *p*-value method. If the sample sizes for the treatments are the same and the observations have a normal distribution with equal variances, then the distribution of $T_{max\ abs}$ may be obtained from the distribution of the studentized range that is the basis for Tukey's multiple comparison procedure. Thus, the minimum *p*-value method may be thought of as a generalization of Tukey's multiple comparison procedure.

SAS Implementation

PROC MULTTEST will perform multiple comparisons based on permutations of the data. Each comparison must be specified separately using a **contrast** statement. The following code may be used, along with the data step from Example 3.2.1 to generate permutation pairwise comparison tests using the minimum *p*-value method.

```
proc multtest data=ta3_2_1 perm ;   /* Perm option requests permutation p-values */
class treat;
contrast 'C vs P1' -1 1 0 0;          /* Contrasts coefficients for each comparison */
contrast 'C vs P2' -1 0 1 0;
contrast 'C vs P3' -1 0 0 1;
contrast 'P1 vs P2' 0 -1 1 0;
contrast 'P1 vs P3' 0 -1 0 1;
contrast 'P2 vs P3' 0 0 -1 1;
test mean(logcount);                  /* Requests t-tests (LSD) for mean differences */
run;
quit;
```

The *p*-values labeled "Permutation" represent *p*-values that have been adjusted by the minimum *p*-value method. The values in the column labeled "Raw" are those obtained using the parametric *t*-test with no adjustment. For instance, the comparison between C and P2 would be significant at the 5% level using the parametric pairwise *t*-test (LSD) but not significant at this level using the minimum *p*-value adjustment.

Output –Section 3.3:

The Multtest Procedure

Model Information

Test for continuous variables:	Mean t-test
Tails for continuous tests:	Two-tailed
Strata weights:	None
P-value adjustment:	Permutation
Center continuous variables?	No
Number of resamples:	20000
Seed:	47374

Contrast Coefficients
treat

Contrast	C	P1	P2	P3
C vs P1	-1	1	0	0
C vs P2	-1	0	1	0
C vs P3	-1	0	0	1
P1 vs P2	0	-1	1	0
P1 vs P3	0	-1	0	1
P2 vs P3	0	0	-1	1

Continuous Variable Tabulations

Variable	treat	NumObs	Mean	Standard Deviation
logcount	C	4	4.1360	0.1302
logcount	P1	6	3.0493	0.5050
logcount	P2	5	3.6690	0.1726
logcount	P3	6	2.7245	0.2818

p-Values

Variable	Contrast	Raw	Permutation
logcount	C vs P1	<.0001	0.0005
logcount	C vs P2	0.0495	0.1899
logcount	C vs P3	<.0001	<.0001
logcount	P1 vs P2	0.0064	0.0288
logcount	P1 vs P3	0.1056	0.3504
logcount	P2 vs P3	0.0002	0.0010

A multiple comparison rank test based on the minimum *p*-value method may be obtained by applying the **PROC MULTTEST** statements above to ranks of the combined observations. An alternative way to rank observations in doing pairwise tests would be to do the rankings separately for each pair of treatments being compared. **PROC MULTTEST** does not allow for this possibility, however.

3.4 Ordered Alternatives

There are situations in which it may be possible to anticipate the direction that the treatments differ. For instance, in a study of the effectiveness of a pain-relieving drug, a researcher may anticipate that the degree of relief will be greater for treatments that use greater levels of the drug. With prior knowledge like this, we can construct statistical tests that are more powerful than the Kruskal-Wallis test and other tests that do not take advantage of such prior knowledge.

Let the null hypothesis be equality of distribution functions of the k treatments. We assume that we have labeled the treatments such that if there is a difference in distributions, we would anticipate that observations from treatment 1 would tend to be smaller than those from treatment 2, and so on. This may be expressed as $H_a: F_1(x) \geq F_2(x) \geq ... \geq F_k(x)$. Let T_{ij} be any test statistic for testing the alternative $H_0: F_i(x) = F_j(x)$ against the *one-sided* alternative $H_a: F_i(x) \geq F_j(x)$, for $i < j$. A general form for the test statistics is the sum of the pairwise statistics T_{ij}, that is, $\sum_{i<j} T_{ij}$.

A reference distribution for testing the above hypotheses can be derived by calculating $\sum_{i<j} T_{ij}$ for each permutation of the observations among the treatments (as described in Section 3.1).

In particular, the Jonckheere-Terpstra (JT) statistic is the sum of pairwise Mann-Whitney statistics. If we let WRS_{ij} denote the Wilcoxon rank sum statistic for comparing treatments i and j where the sum of the ranks is taken over the observations in treatment j, $i < j$, then the Mann-Whitney statistic for comparing treatments i and j is $MW_{ij} = WRS_{ij} - (n_{ij})(n_j + 1)/2$. See Higgins (2004) for a discussion of the Mann-Whitney statistic. Then $JT = \sum_{i<j} MW_{ij}$ can be computed for each permutation to obtain an exact p-value. Alternatively, an asymptotic approximation is based on the approximate normality of $Z = (JT - E(JT))/\sqrt{Var(JT)}$. See Lehmann (1975) for formulas for the mean and variance of the JT statistic.

SAS Implementation
PROC FREQ can be used to perform the Jonckheere-Terpstra test.

Example 3.4.1. An agronomist studied the effect of mowing height on the phosphorous content of a certain species of prairie grass. He postulated that phosphorous levels would tend to be lower in plants that have been mowed at greater heights.

Table 3.4.1. Phosphorous contents of plants under four mowing treatments.

Unmowed:	4.302	4.017	4.049	4.176		
20 cm:	2.021	3.190	3.250	3.276	3.292	3.267
10 cm:	3.397	3.552	3.630	3.578	3.612	
5 cm:	2.699	2.929	2.785	2.176	2.845	2.913

The following code performs the Jonckheere-Terpstra test on the data of Table 3.4.1.

```
data ta3_4_1;
input trtcode treat $ phos @@;
datalines;
1 un   13      1 un   24.1   1 un   11.7   1 un   16.3   1 un   15.5   1 un   24.5
2 20   42      2 20   18      2 20   14      2 20   36      2 20   11.6   2 20   19
3 10   15.6    3 10   23.8    3 10   24.4   3 10   24      3 10   21     3 10   21.1
4 5    35.3    4 5    22.5    4 5    16.9    4 5    25      4 5    23.1   4 5    26
;

proc freq data=ta3_4_1;
tables trtcode*phos/jt noprint;        /* Requests JT test,  suppress printing frequency tables */
exact jt;                              /* Requests exact p-value */
run;
```

The one-sided *p*-value of 0.0285, based on the large sample approximation is given in **(1)**, while the exact *p*-value, 0.0299, is given in **(2)**.

Output – Example 3.4.1:

```
              The FREQ Procedure

        Statistics for Table of trtcode by phos

              Jonckheere-Terpstra Test

          Statistic (JT)           145.0000
          Z                          1.9031

(1)       Asymptotic Test
          One-sided Pr > Z           0.0285
          Two-sided Pr > |Z|         0.0570

(2)       Exact Test
          One-sided Pr >=  JT        0.0299
          Two-sided Pr >= |JT - Mean|  0.0598

              Sample Size = 24
```

Caution: It is necessary that the values of the classification variable be ordered in the same way as the hypothesized orderings of the populations. Notice in the example above that the variable "trtcode" was used instead of the variable "treat". If the actual values of the variable treat were used, the following (incorrect) results would result:

Output –Example 3.4.1 (continued):

```
                       The FREQ Procedure

              Statistics for Table of treat by phos

                    Jonckheere-Terpstra Test
          ─────────────────────────────────────────────
          Statistic (JT)               93.0000
          Z                            -0.7715

          Asymptotic Test
          One-sided Pr <  Z             0.2202
          Two-sided Pr > |Z|            0.4404

          Exact Test
          One-sided Pr <=  JT           0.2303
          Two-sided Pr >= |JT - Mean|   0.4606

                     Sample Size = 24
```

The problem is that because the classification variable is a string variable, the treatments are ordered alphabetically, resulting in the ordering, from assumed smallest to largest: 10, 20, 5, un. The new variable "trtcode" correctly reflects the hypothesized ordering of the distributions.

Analysis of the same data using the Kruskal-Wallis test, which does not take advantage of the ordering among alternatives, shows no statistically significant difference among the treatments. The estimated *p*-value is 0.3344 based on 20,000 randomly selected permutations, as shown below.

Output – Example 3.4.1 (continued):

```
                      The NPAR1WAY Procedure

           Wilcoxon Scores (Rank Sums) for Variable phos
                    Classified by Variable treat

                        Sum of    Expected    Std Dev       Mean
        treat    N      Scores    Under HO    Under HO      Score
        ───────────────────────────────────────────────────────────
        un       6       53.0       75.0        15.0      8.833333
        20       6       71.0       75.0        15.0     11.833333
        10       6       78.0       75.0        15.0     13.000000
        5        6       98.0       75.0        15.0     16.333333


                        Kruskal-Wallis Test

               Chi-Square            3.4600
               DF                         3
               Pr > Chi-Square       0.3260


             Monte Carlo Estimate for the Exact Test

                  Pr >= Chi-Square
                  Estimate              0.3344
                  99% Lower Conf Limit  0.3258
                  99% Upper Conf Limit  0.3429

                  Number of Samples      20000
                  Initial Seed           69894
```

4

Paired Comparisons and Blocked Designs

4.1 Paired Comparisons

Pairing and blocking are experimental design techniques that enable a researcher to detect differences among treatments more easily in environments where there is a lot of variability among experimental units. For instance, suppose we wish to compare two brands of a food product for taste. One way to do the experiment would be to have one group of consumers taste product A and another group taste product B. Consumers in the experiment would be assigned randomly to either product A or B. This is a two-sample experiment for which the techniques of Chapter 2 would be appropriate. On the other hand, we could have each consumer taste both brands, where the order of tasting is determined randomly. This experiment is a paired-comparison experiment. Intuitively, it ought to be better for determining brand preference because each consumer has a chance to evaluate both brands.

The analysis of paired-comparison data is done on differences. Suppose X_i and Y_i are the observations on treatments A and B, respectively for the ith pair, $i = 1, 2, ..., n$, and let $D_i = X_i - Y_i$. The one-sample t-test may be applied to the differences if they comprise a random sample from a normal distribution. A permutation test for paired comparisons may be used if the normality assumption is questionable. The permutation test is based on the fact that X_i and Y_i are interchangeable within the pair if there is no difference between treatments; that is, it is as likely that X_i appears with B and Y_i appears with A as the other way around. In this case, the differences may take on the values $\pm|D_i|$ with probability 0.5 each, and the set of permutations consists of the 2^n assignments of plus and minus signs to $|D_i|$. Various equivalent permutation statistics may be used in testing for differences between treatments. These include the sum of positive differences, the sum of negative differences, and the sum of all differences.

Scores may be attached to the differences, thereby producing various tests. If the scores are the signs of the differences, then the permutation test is the sign test. If the scores are the ranks of the absolute values of the differences, then the permutation test is the Wilcoxon signed-rank test. SAS ignores differences that are zero. For instance, if there are 10 differences, and one of them is zero, then only the nonzero differences would be ranked in doing the signed-rank test. Another version of the signed-rank test includes the zeros in the ranking of the absolute values of the observations. This is discussed in Lehmann (1975) but is not implemented in SAS.

SAS Implementation

The **UNIVARIATE** procedure will compute exact p-values for the sign and Wilcoxon signed-ranks tests. An exact p-value based on the permutation distribution of the raw data is not available.

Example 4.1.1. Seventeen pairs of twins were involved in a study of cholesterol-reducing drugs. One of the twins in each pair was given drug 1 and the other was given drug 2, where the choice was made at random. The amount by which the cholesterol was reduced in each subject is shown in Table 4.1.1.

Table 4.1.1. Cholesterol reduction for two drugs.						
Pair	Drug 1	Drug 2	Difference	Pair	Drug 1	Drug 2	Difference
1	74	63	11	10	58	38	20
2	55	58	-3	11	54	56	-2
3	61	49	12	12	53	38	15
4	41	47	-6	13	69	47	22
5	53	50	3	14	60	41	19
6	74	69	5	15	61	46	15
7	52	67	-15	16	54	47	7
8	31	40	-9	17	57	44	13
9	50	44	6				

The following code generates both the sign test and Wilcoxon signed-rank test, as well as the parametric *t*-test for the data of Table 4.1. Note that the statistic used by SAS for the sign test is $M = (SN_+ - SN_-)/2$, where SN_+ is the number of positive differences and SN_- is the number of negative differences. M is equivalent to using either of SN_+ or SN_- as the test statistic. For both the sign test and the signed-rank test, observations with zero differences are omitted from the analysis.

```
data ta4_1_1;
input pair drug1 drug2 @@;
diff=drug1-drug2;          /* Creates difference variable */
datalines;
1  74 63  2  55 58  3  61 49  4  41 47  5  53 50  6  74 69
7  52 67  8  31 40  9  50 44 10 58 38 11 54 56 12 53 38
13 69 47 14 60 41 15 61 46 16 54 47 17 57 44
;

proc univariate data=ta4_1_1;
var diff;
run;
```

The two-sided *p*-values are 0.0215 for the parametric *t*-test (**1**), 0.1435 for the sign test (**2**), and 0.0256 for the signed-rank test (**3**).

Output – Example 4.1.1:

```
                        The UNIVARIATE Procedure
                            Variable:  diff

                               Moments

        N                    17    Sum Weights               17
        Mean            6.64705882    Sum Observations          113
        Std Deviation   10.7583757    Variance             115.742647
        Skewness        -0.4580626    Kurtosis              -0.6473968
        Uncorrected SS        2603    Corrected SS         1851.88235
        Coeff Variation  161.851669    Std Error Mean       2.60928937


                      Basic Statistical Measures

            Location                        Variability

        Mean       6.64706      Std Deviation        10.75838
        Median     7.00000      Variance            115.74265
        Mode      15.00000      Range                37.00000
                                Interquartile Range  17.00000


                    Tests for Location: Mu0=0

            Test          -Statistic-      -----p Value------
```

	Test	-Statistic-	-----p Value------	
(1)	Student's t	t 2.547459	Pr > |t|	0.0215
(2)	Sign	M 3.5	Pr >= |M|	0.1435
(3)	Signed Rank	S 46.5	Pr >= |S|	0.0256

4.2   Friedman's Test for a Randomized Complete Block Design

The randomized complete block design extends the paired-comparison design to more than two treatments. Within blocks the experimental units are more homogenous than between blocks. Each of the k treatments appears exactly once in each block with units being randomly assigned to treatments within blocks. In agronomy, the blocks are typically plots of ground, and the experimental units are subdivisions of the plots.

Let X_{ij} denote the jth observation in the ith block, $i = 1, 2, ..., k$, $j = 1, 2, ..., b$. The randomized complete block model is $X_{ij} = \mu + t_i + b_j + \varepsilon_{ij}$, where μ is an overall effect, t_i is the effect of the ith treatment, b_j is the effect of the jth block, and the ε_{ij}'s are independent, identically distributed random variables with a median of 0. Let the row, column, and overall means be given by $\bar{X}_{i.} = \frac{1}{b} \sum_{j=1}^{b} X_{ij}$, $\bar{X}_{.j} = \frac{1}{k} \sum_{i=1}^{k} X_{ij}$, and $\bar{X} = \frac{1}{bk} \sum_{i=1}^{k} \sum_{j=1}^{b} X_{ij}$. The total sum of squares is

$$SS_{total} = \sum_{i=1}^{k} \sum_{j=1}^{b} \left(X_{ij} - \bar{X} \right)^2.$$ The sum of squares for treatments, blocks, and error are

$$SST = b \sum_{i=1}^{k} \left(\bar{X}_{i.} - \bar{X} \right)^2, \quad SSB = k \sum_{j=1}^{b} \left(\bar{X}_{.j} - \bar{X} \right)^2, \text{ and } SSE = \sum_{i=1}^{k} \sum_{j=1}^{b} \left(X_{ij} - \bar{X}_{i.} - \bar{X}_{.j} + \bar{X} \right)^2, \text{ respectively.}$$

If the ε_{ij}'s have a normal distribution, then a test for significant treatment effect may be carried out

with the statistic $F = \dfrac{SST/(k-1)}{SSE/[(k-1)(b-1)]}$. The reference distribution is the F-distribution with $k -$
1 degrees of freedom for the numerator and $(k-1)(b-1)$ degrees of freedom for the denominator.

     If the normality assumption is not valid, then we may carry out a permutation F-test, where observations are permuted within blocks. With each permutation of the data, F is computed. The permutation p-value is the fraction of the F's greater than or equal to the observed value F_{obs}. There are various statistics that are equivalent to the permutation F-statistic. We note, for instance, that $SSE = SS_{total} - SSB - SST$. Both SS_{total} and SSB are invariant with respect to permutations of the data. Thus, the permutation F-statistic may be expressed as a one-to-one function of SST, and so, F and SST are equivalent permutation statistics.

     A permutation test applied to ranks, where ranking is done within blocks, is equivalent to Friedman's test. The usual form of Friedman's statistic is $cSST_{ranks}$, where SST_{ranks} is the sum of squares for the treatments applied to ranks, and c is chosen so that the asymptotic distribution is chi-square with $k-1$ degrees of freedom. If there are no ties among the ranks, then

$$c = 12/[k(k+1)].$$ If there are ties, then $c = b/\sum_{j=1}^{b} S_{B_j}^2$ where $S_{B_j}^2$ is the sample variance of the

scores for ranks of the observations in the jth block adjusted for ties if necessary. This form of the statistic may be used for other scores as well. The asymptotic chi-square approximation works well for this test even for relatively small sample sizes. If there are just two treatments, then Friedman's test is equivalent to the sign test. Generally, Friedman's test will have low power unless there are a large number of blocks.

SAS Implementation

PROC FREQ can be used to produce Friedman's test.

Example 4.2.1. Table 4.2.1 gives hypothetical data arising from an experiment to compare four treatments with experimental units arranged in a randomized complete block design.

Table 4.2.1. Hypothetical data, randomized complete block design with four treatments.

Treatment	Block 1	Block 2	Block 3
1	100	80	50
2	100	80	60
3	150	80	80
4	150	90	90

The following code generates the large sample approximation for Friedman's test, using the data of Table 4.2.1.

```
data ta4_2_1;
input treat block resp @@;
datalines;
1 1 100 1 2 80 1 3 50
2 1 100 2 2 80 2 3 60
3 1 150 3 2 80 3 3 80
4 1 150 4 2 90 4 3 90
;
proc freq data=ta4_2_1;
tables block*treat*resp/
       cmh2 scores=rank        /* Requests Friedman statistic */
       noprint;                /* Suppresses printing of frequency tables */
run;
```

The Friedman statistic is a special case of the Cochran-Mantel-Haenszel (CMH) statistic. When rank scores are used, the entry of CMH for "Row Mean Scores Differ" is identical to the Friedman statistic. Thus, the combination of the **cmh2** and **scores=rank** options produce the Friedman statistic. For this example, the value of the Friedman statistic is 7.5 and the corresponding *p*-value, based on the large sample approximation, is 0.0576 **(1)**. The **noprint** option suppresses printing of the frequency tables that are generated by default using the **tables** statement.

Output – Example 4.2.1:

```
                          The FREQ Procedure

                  Summary Statistics for treat by resp
                         Controlling for block

           Cochran-Mantel-Haenszel Statistics (Based on Rank Scores)

       Statistic   Alternative Hypothesis    DF    Value    Prob
       ─────────────────────────────────────────────────────────
          1        Nonzero Correlation        1    7.2000   0.0073
 (1)      2        Row Mean Scores Differ     3    7.5000   0.0576


                    Total Sample Size = 12
```

5

Tests of Association for Bivariate Data

5.1 Measures of Association for Bivariate Data

Let $(X_i, Y)_i$, $i = 1,...,n$, denote n pairs of random variables such as the heights and weights of n individuals. Of interest are statistics that measure the tendency of the $Y's$ to increase or decrease with an increase or decrease in the $X's$.

The *Pearson product-moment correlation* is defined as

$$r = \sum_{i=1}^n (X_i - \bar{X})(Y_i - \bar{Y}) \Big/ \sqrt{\sum_{i=1}^n (X_i - \bar{X})^2 \sum_{i=1}^n (Y_i - \bar{Y})^2} \,.$$

It measures the strength of linear relationship between the $X's$ and $Y's$. If the $X's$ and $Y's$ have a bivariate normal distribution with population correlation ρ, then the hypothesis $H_0 : \rho = 0$ may be tested against one-sided or two-sided alternatives using the statistic $t = \sqrt{\dfrac{n-2}{1-r^2}}\, r$, which has a t-distribution with $n-2$ degrees of freedom. This test is equivalent to the test that the slope is zero in simple linear regression under the assumption of normally distributed errors.

We may also carry out a permutation test, which does not require the assumption of normality. The permutation distribution of r may be obtained by pairing the $Y's$ with the $X's$ in all $n!$ possible ways and computing r for each pairing. The exact p-value for an observed statistic is the fraction of the permutation distribution as extreme or more extreme than the observed statistic. If it is not feasible to obtain all permutations, then a random sample of them may be used to obtain an approximate p-value.

The *Spearman rank correlation*, r_s, is the Pearson product moment correlation computed using the ranks of the $X's$ and the ranks of the $Y's$. It measures the extent to which the $X's$ and $Y's$ are in the same rank order, and thus the strength of the monotonic relationship. The permutation test can be used to test the null hypothesis of no association between the ranks.

Under the null hypothesis of no association, the permutation distributions of both r and r_s have mean 0 and variance $1/(n-1)$, and the permutation distributions of the standardized correlations have approximate standard normal distributions. See Higgins (2004, Chapter 5). Equivalently, the permutation distributions of the quantities $(n-1)r^2$ and $(n-1)r_s^2$ have approximate chi-square distributions with 1 degree of freedom under the hypothesis of no association. We may obtain asymptotic two-sided p-values for the permutation tests from the chi-square distribution (See Section 5.2).

Other common measures of association are based on the notion of *concordance* and *discordance* between bivariate pairs. A pair of bivariate points $(X_i, Y_i), (X_j, Y_j)$ is said to be *concordant* if $(X_i - X_j)(Y_i - Y_j) > 0$, that is, if both coordinates either increase or decrease. If the product is negative, then the pair is said to be *discordant;* if the product is 0, then the ordering of the pair is indeterminate. Let C denote the number of concordant pairs and D denote the number of discordant pairs among all pairs of points.

The *gamma coefficient* is computed as $r_\gamma = \dfrac{C-D}{C+D}$. If there are ties among either the X or Y values, then the gamma value reported by SAS is computed as above with any pairs containing ties on either X or Y ignored.

Kendall's tau-b is computed as $r_{\tau_b} = \dfrac{C-D}{\sqrt{\left[\binom{n}{2}-T_X\right]\left[\binom{n}{2}-T_Y\right]}}$, where T_X is the number of pairs that

are tied in the X coordinate only and T_Y is the number of pairs that are tied in the Y coordinate only.

Somers' D is similar to Kendall's tau-b but omits either only pairs tied on X, $r_{S_x} = \dfrac{C-D}{\binom{n}{2}-T_X}$, or

those tied on Y, $r_{S_y} = \dfrac{C-D}{\binom{n}{2}-T_Y}$ (thus, two measures are generated).

When there are no ties on either the X or Y values, then all of these measures will be the same.

Another less common measure of association based on concordance is the *Hoeffding statistic*. It uses the fact that the quantity $P(X \le x, Y \le y) - P(X \le x)P(Y \le y)$ is a measure of the departure of the joint distribution from independence. The Hoeffding statistic is an estimate of the expected square of this quantity.

All of the above measures, with the exception of the Pearson product-moment correlation, are rank-based and may be applied to bivariate ordinal data. For instance, we may apply such measures to leaf damage and root damage of a plant using the ordinal responses none, small, moderate, and severe to denote the level of damage. Exact p-values for tests of each of these measures of association may be obtained from their respective permutation distributions. The asymptotic permutation distributions are normal except for that of the Hoeffding statistic. Formulas for the permutation variance and a discussion of the Hoeffding statistic may be found in Hollander and Wolfe (1999, Chapter 8).

Most of these statistics can be obtained in more than one way in SAS. In some cases, exact p-values for testing the null hypothesis of no association between X and Y may be requested; in other cases, only asymptotic p-values are available. SAS uses different asymptotic approximations depending on the statistic and the procedure. For instance, there are three approximations for the asymptotic p-value of the Spearman correlation r_S. The approximation implemented in **PROC**

CORR uses the statistic $t = \sqrt{\dfrac{n-2}{1-r_S^2}}\, r_S$, which is assumed to have a t-distribution with $n-2$

degrees of freedom. Another approximation implemented in **PROC FREQ** uses the results of Brown and Benedetti (1977) for multinomial sampling. Here the p-value is obtained from the approximately normally distributed statistic $Z = r_S / \sqrt{V_{mult}(r_S)}$, where $V_{mult}(r_S)$ is the variance of r_S under the assumption of multinomial sampling. A third is the chi-square approximation of the permutation distribution of $(n-1)r_S^2$, which is also available in **PROC FREQ**. In keeping with the nonparametric theme of this book, our preference is to use exact tests, based on the permutation distributions of the statistics, where available, and the large sample approximations of these when the exact test is not available.

SAS Implementation

Example 5.1.1. Table 5.1.1 gives a hypothetical set of bivariate observations that will be used to illustrate the different SAS procedures for obtaining the measures of association described above.

Table 5.1.1. Hypothetical set of six bivariate observations.

X	1.2	2.4	4.0	4.0	5.4	6.3
Y	3.5	2.9	1.6	5.9	4.7	6.7

The following code illustrates the different ways to generate estimates and tests for the different measures of association using the data of Table 5.1.1.

```
data ta5_1_1;
input x y @@;
datalines;
1.2 3.5 2.4 2.9 4.0 1.6 4.0 5.9 5.4 4.7 6.3 6.7
;

proc freq data=ta5_1_1;
tables x*y /
        measures            /* Requests estimates and tests for association measures */
        jt                  /* Requests Jonckheere-Terpstra test */
        chisq               /* Requests chi-square tests */
        cmh                 /* Requests Cochran-Mantel-Haenszel tests */
        noprint;            /* Suppresses print of frequency tables */
exact measures jt chisq;    /* Requests exact tests */
run;

proc freq data=ta5_1_1;
tables x*y /
        measures            /* Requests estimates and tests for association measures */
        chisq               /* Requests chi-square tests */
        cmh                 /* Requests Cochran-Mantel-Haenszel tests */
        scores=rank         /* Requests analysis on ranks */
        noprint;            /* Suppresses print of frequency tables */
exact measures chisq;
run;

proc corr data=ta5_1_1 pearson spearman kendall hoeffding;
var x y;
run;
```

The previous code generates all (and more) of the following output, although the output has been edited and rearranged for ease of presentation. Sections **(1)**, **(2)**, **(3)** and **(4)** are produced by **PROC FREQ**. Section **(1)** is generated from the **measures** option in the **tables** statement and gives the estimates for the association measures discussed previously, with the exception of the Hoeffding statistic. Sections **(2)**, **(3)**, and **(4)** give tests for the Pearson correlation. Section **(2)** is generated by the **measures** option in the **exact** statement. The exact two-sided p-value is 0.2333. The asymptotic p-value given in **(2)** is not the large sample approximation to the permutation p-value but rather is based on the multinomial sampling assumption of Brown and Benedetti (1977). Section **(3)** is generated by the **chisq** option in the **exact** statement. The statistic is $(n-1)r^2 = 1.7125$, and thus the exact p-value 0.2333 is the same as given in Section **(2)**. The asymptotic p-value of 0.1907 is based on the chi-square approximation to the permutation distribution. This value is also given in Section **(4)**, which is generated by the **cmh** option in the **tables** statement. Section **(5)** is generated by **PROC CORR**, and the p-value is based on the statistic $t = \sqrt{\dfrac{n-2}{1-r^2}}\, r$, assumed to have a t-distribution with $n-2$ degrees of freedom.

Output – Example 5.1.1:

(1)

```
                    Statistics for Table of x by y

      Statistic                            Value      ASE
      -------------------------------------------------------
      Gamma                                0.4286    0.2723
      Kendall's Tau-b                      0.4140    0.2691
      Stuart's Tau-c                       0.4167    0.2778

      Somers' D C|R                        0.4286    0.2723
      Somers' D R|C                        0.4000    0.2667

      Pearson Correlation                  0.5852    0.1972
      Spearman Correlation                 0.6088    0.2769

      Lambda Asymmetric C|R                0.8000    0.1789
      Lambda Asymmetric R|C                1.0000    0.0000
      Lambda Symmetric                     0.8889    0.1104

      Uncertainty Coefficient C|R          0.8710    0.0744
      Uncertainty Coefficient R|C          1.0000    0.0000
      Uncertainty Coefficient Symmetric    0.9311    0.0425
```

(2)

```
               Pearson Correlation Coefficient
               --------------------------------
               Correlation (r)          0.5852
               ASE                      0.1972
               95% Lower Conf Limit     0.1988
               95% Upper Conf Limit     0.9717

               Test of H0: Correlation = 0

               ASE under H0             0.2832
               Z                        2.0668
               One-sided Pr >  Z        0.0194
               Two-sided Pr > |Z|       0.0388

               Exact Test
               One-sided Pr >=  r       0.1167
               Two-sided Pr >= |r|      0.2333
```

Output – Example 5.1.1 (continued):

(3)
```
                          Mantel-Haenszel Chi-Square Test
              ---------------------------------
              Chi-Square                1.7125
              DF                             1
              Asymptotic Pr >  ChiSq    0.1907
              Exact       Pr >= ChiSq   0.2333
```

(4)
```
              Cochran-Mantel-Haenszel Statistics (Based on Table Scores)

      Statistic   Alternative Hypothesis   DF     Value    Prob
      ------------------------------------------------------------
          1       Nonzero Correlation       1     1.7125   0.1907
          2       Row Mean Scores Differ    4     2.4779   0.6486
          3       General Association      20    20.0000   0.4579
```

(5)
```
                  Pearson Correlation Coefficients, N = 6
                     Prob > |r| under HO: Rho=0


                               x              y

                  x        1.00000        0.58524
                                          0.2224


                  y        0.58524        1.00000
                           0.2224
```

Similar output is generated for Spearman correlation in Sections **(6)**, **(7)**, **(8)**, and **(9)**. Sections **(6)**, **(7)**, and **(8)** result from using the **scores=rank** option in the **tables** statement. The exact two-sided p-value is 0.2222, and the large sample approximation is 0.1734.

Output – Example 5.1.1 (continued):

(6)
```
                          Spearman Correlation Coefficient
              --------------------------------
              Correlation (r)           0.6088
              ASE                       0.2769
              95% Lower Conf Limit      0.0661
              95% Upper Conf Limit      1.0000

                  Test of HO: Correlation = 0

              ASE under HO              0.2848
              Z                         2.1377
              One-sided Pr >  Z         0.0163
              Two-sided Pr > |Z|        0.0325

              Exact Test
              One-sided Pr >=  r        0.1111
              Two-sided Pr >= |r|       0.2222
```

Output – Example 5.1.1 (continued):

(7)

```
              Mantel-Haenszel Chi-Square Test
                     (Rank Scores)
         -----------------------------------
         Chi-Square                   1.8529
         DF                                1
         Asymptotic Pr >  ChiSq       0.1734
         Exact      Pr >= ChiSq       0.2222
```

(8)

```
        Cochran-Mantel-Haenszel Statistics (Based on Rank Scores)
```

Statistic	Alternative Hypothesis	DF	Value	Prob
1	Nonzero Correlation	1	1.8529	0.1734
2	Row Mean Scores Differ	4	2.7143	0.6067
3	General Association	20	20.0000	0.4579

(9)

```
              Spearman Correlation Coefficients, N = 6
                   Prob > |r| under H0: Rho=0
```

	x	y
x	1.00000	0.60876
		0.1997
y	0.60876	1.00000
	0.1997	

Sections **(10)** and **(11)** are produced by **PROC FREQ**. An exact *p*-value is not available for Kendall's tau-*b* or the other measures of concordance. However, the Jonckheere-Terpstra test (see Section 3.4) is equivalent to the test of no association for Kendall's tau-*b*, and an exact *p*-value is available for the Jonckheere-Terpstra test. Section **(11)** is generated by the **jt** option in the **exact** statement, yielding an exact two-sided *p*-value of 0.3556, with large sample approximation 0.2511. The large sample approximation is also generated by **PROC CORR** in Section **(12)**. Finally, Section **(13)**, generated by **PROC CORR**, gives the Hoeffding coefficient of dependence $D = 0.0000$, with asymptotic *p*-value 0.3643.

Output – Example 5.1.1 (continued):

(10)

```
                     Kendall's Tau-b
         -------------------------------
         Tau-b                        0.4140
         ASE                          0.2691
         95% Lower Conf Limit        -0.1133
         95% Upper Conf Limit         0.9414

              Test of H0: Tau-b = 0

         ASE under H0                 0.2760
         Z                            1.5000
         One-sided Pr >  Z            0.0668
         Two-sided Pr > |Z|           0.1336
```

Output – Example 5.1.1 (continued)**:**

(11)

```
                          Jonckheere-Terpstra Test
              -----------------------------------
              Statistic (JT)              10.0000
              Z                            1.1476

              Asymptotic Test
              One-sided Pr >  Z            0.1256
              Two-sided Pr > |Z|           0.2511

              Exact Test
              One-sided Pr >=  JT          0.1778
              Two-sided Pr >= |JT - Mean|  0.3556
```

(12)

```
              Kendall Tau b Correlation Coefficients, N = 6
                       Prob > |r| under H0: Rho=0

                            x               y

              x         1.00000         0.41404
                                        0.2511

              y         0.41404         1.00000
                        0.2511
```

(13)

```
                          The CORR Procedure

              Hoeffding Dependence Coefficients, N = 6
                       Prob > D under H0: D=0

                            x               y

              x         0.60938         0.00000
                        0.0010          0.3643

              y         0.00000         1.00000
                        0.3643
```

Table 5.1.2 gives a summary of the various *p*-values produced for each of the measures in the previous example.

Table 5.1.2. Summary of *p*-value comparisons for tests of no association. Asterisks indicate large sample approximations to the permutation distribution. The *p*-value given for the **exact** option is the exact permutation *p*-value.

Procedure	Statistic	Option	p-value
FREQ	Pearson	**measures**	0.0388
		chisq	0.1907*
		cmh	0.1907*
		exact	**0.2333**
	Spearman	**measures**	0.0325
		chisq scores=ranks	0.1734*
		cmh scores=ranks	0.1734*
		exact	**0.2222**
	Kendall tau-b	**measures**	0.1336
		exact jt	0.2511*/0.3556(exact)
CORR	Pearson	**pearson**	0.2224
	Spearman	**spearman**	0.1997
	Kendall tau-b	**kendall**	0.2511*
	Hoeffding	**hoeffding**	0.3643*

5.2 Summary of Options for Obtaining *P*-values for Measures of Association

Tables 5.2.1 and 5.2.2 summarize the options that will produce the tests for the various measures of association, whether or not exact tests are available and the type of asymptotic approximation that is used. These are denoted *perm* for the approximation based on permutation tests, and *mult* for the approximation based on multinomial sampling. The **chisq** and **cmh** options give the chi-square statistic and the two-sided significance levels, but not the actual measure of association. The outputs are labeled "MH Chi-Square" for the **chisq** option and "Nonzero Correlation" for the **cmh** option. These are discussed in more detail in Chapter 6.

Table 5.2.1. Options for measures of association with **PROC FREQ**.

Statistic	Tables Options	Exact Test Available?	Asymptotic Approximation
Pearson	**measures**	yes	perm
	chisq	yes	mult
	cmh	no	perm
Spearman	**measures**	yes	perm
	chisq, scores=ranks	yes	mult
	cmh, scores=ranks	no	perm
Kendall, Kendall-like	**measures**	no[*]	mult

* Exact *p*-value available using **jt** option.

Table 5.2.2. Options for measures of association with **PROC CORR**.

Statistic	CORR options	Exact Test Available	Asymptotic Approximation
Pearson	**pearson**	no	t-test
Spearman	**spearman**	no	t-test
Kendall	**kendall**	no	perm
Hoeffding	**hoeffding**	no	perm

6

Tests of Association for Contingency Tables

6.1 Measures of Association for Contingency Tables

We consider measuring the association between the two factors in an $R \times C$ contingency table. Let n_{ij} denote the count in cell (i, j) of the contingency table, $i = 1, ..., R$ and $j = 1, ..., C$. Let $n_{i.}$ and $n_{.j}$ denote the totals for row i and column j, respectively, and let n denote the total number of observations. The expected value for cell (i, j) under the assumption of no association is given by $e_{ij} = \dfrac{n_{i.}n_{.j}}{n}$. SAS computes the *Pearson chi-square* $\left(\chi^2\right)$ *statistic* for a contingency table and three measures of association based on χ^2.

The Pearson chi-square statistic for a contingency table is $\chi^2 = \sum\limits_{i,j} \dfrac{\left(n_{ij} - e_{ij}\right)^2}{e_{ij}}$.

For 2×2 tables, the *phi coefficient* is computed as $\phi = \dfrac{n_{11}n_{22} - n_{12}n_{21}}{\sqrt{n_{1.}n_{2.}n_{.1}n_{.2}}}$, and for larger tables as $\phi = \sqrt{\dfrac{\chi^2}{n}}$.

The *contingency coefficient* is computed as $CC = \sqrt{\dfrac{\chi^2}{\chi^2 + n}}$.

For $R \times C$ tables, *Cramer's V* is given by $V = \sqrt{\dfrac{\chi^2/n}{\min\left(R-1, C-1\right)}}$. For 2×2 tables, $V = \phi$.

The *likelihood ratio chi-square statistic* is $LR = -2\log(L)$, where L is the likelihood ratio statistic given by $L = \prod\limits_{i,j} \left(\dfrac{e_{ij}}{n_{ij}}\right)^{n_{ij}}$.

The *Mantel-Haenszel chi-square statistic* measures association between the row and column factors. If the levels of both factors are numerical, then the statistic is $MH = (n-1)r^2$, where r is the Pearson product-moment correlation. If a factor is ordinal but non-numeric, SAS uses the numerical value 1 for the category label with the lowest alpha-numeric ordering, 2 for the category label with the next lowest alpha-numeric ordering, and so on in computing the correlation.

numerical value 1 for the category label with the lowest alpha-numeric ordering, 2 for the category label with the next lowest alpha-numeric ordering, and so on in computing the correlation.

6.2 Chi-Square and Permutation Tests for Contingency Tables

The Pearson chi-square statistic and $-2\log L$ have asymptotic chi-square distributions with degrees of freedom $(R-1)(C-1)$. The Mantel-Haenszel chi-square statistic has an asymptotic chi-square distribution with 1 degree of freedom. In cases where cell counts are small, the chi-square approximations may not be sufficiently accurate. Alternatively, we may obtain an exact permutation p-value for the statistic as outlined below. Example 6.2.1 shows how a permutation test is carried out for a contingency table.

Example 6.2.1. Seven patients were selected for a study to compare two methods of relieving postoperative pain. Three were allowed to control the amount of pain-relief medicine themselves according to their level of pain. The other four were given a physician-prescribed level of medicine. Afterward the patients were asked to evaluate the effectiveness of their pain-relief treatment, with responses being not satisfied, somewhat satisfied, and very satisfied. The data are given in Table 6.2.1.

Table 6.2.1. Satisfaction with pain relief treatment.			
	Not Satisfied NS	Somewhat Satisfied SS	Very Satisfied VS
Physician-prescribed (PP)	2	2	0
Self-administered (SA)	0	1	2

Label the treatments and responses of the seven patients as (A_i, B_i), where the subscript denotes the patient number, $i = 1,...,7$; A_i denotes the row label for the ith patient, which can take one of the values PP or SA; and B_i denotes the column label for the ith patient, which can take on one of the values NS, SS, or VS. There are three ways to permute the data that will yield the same permutation p-value for any of the measures of association, assuming that the marginal totals are fixed. First, we may permute the B_i's among the A_i's in the 7! possible ways, as in the permutation test for bivariate association described in Section 5.1. Alternatively, we may permute the 7 column labels NS, NS, SS, SS, SS, VS, VS between the 2 rows in the $\binom{7}{4}$ ways, as in the two-sample permutation test described in Section 2.1. Finally, we may permute the 7 row labels PP, PP, PP, PP, SA, SA, SA among the 3 columns in the $\frac{7!}{2!3!2!}$ ways, as in the k-sample permutation test described in Section 3.1. For each of the possible permutations, we compute the statistic of interest. This set of values forms the permutation distribution from which the p-value of the observed statistic may be obtained. For instance, if the statistic is the Pearson chi-square, then the p-value is the fraction of the chi-square values from the permutation distribution that are greater than or equal to the observed value.

SAS Implementation

The following code generates chi-square statistics, with both asymptotic and exact (permutation) p-values, for the data of Table 6.2.1. Also displayed are the three measures of association for nominal data discussed in Section 6.1. The first program gives an example of data entered as individual observations, and the second as data entered as a frequency table. In the latter case, a **weight** statement is used to identify the count variable. If the labels of the rows and columns are non-numeric, then SAS will use the alpha-numeric ordering of the labels when computing the Mantel-Haenszel statistic. In the two ways that the data are entered below, both the numeric and non-numeric labels are in the same order, so both programs yield identical output. Ordering of the variable labels does not affect the computation of the Pearson or likelihood ratio chi-square statistics.

```
/* Program 1: Data entered as individual observations */
data ta6_2_1;
input treat $ satisf $ @@;
datalines;
pp ns pp ns pp ss pp ss sa ss sa vs sa vs
;

proc freq data=ta6_2_1;
tables treat*satisf;
exact chisq;                    /* Requests exact and asymptotic chi-square tests */
run;
```

```
/* Program 2: Data entered as frequencies */
data ta6_2_1;
input treat satisf count @@;
datalines;
1 1 2 1 2 2 1 3 0
2 1 0 2 2 1 2 3 2
;

proc freq data=ta6_2_1;
weight count;                   /* Indicates frequency variable */
exact chisq;                    /* Requests exact and asymptotic chi-square tests */
tables treat*satisf;
run;
```

Section **(1)** gives the three chi-square statistics and along with asymptotic p-values for tests of no association, and also gives the three measures of association related to the Pearson chi-square statistic. Note that the Mantel-Haenszel chi-square test is printed but has no meaning for this example because the data are nominal and there is no real ordering to the category labels.

Below Table **(1)** are separate tables displaying exact and asymptotic p-values for each of the chi-square statistics. The Pearson chi-square statistic is 4.2778, and the likelihood ratio chi-square statistic is 5.7416. The exact p-value is 0.3143 for both statistics, while the asymptotic p-values are different (0.1178 and 0.0567, respectively). Note the large difference between the exact and asymptotic p-values in this case. Because the cell sample sizes are all very small, the asymptotic p-values based on the chi-square distribution do not do a good job of approximating the exact value.

Output – Example 6.2.1:

```
                        The FREQ Procedure

                   Table of treat by satisf

        treat     satisf

        Frequency|
        Percent  |
        Row Pct  |
        Col Pct  |ns      |ss      |vs      | Total
        ---------+--------+--------+--------+
        pp       |    2 |     2 |     0 |      4
                 | 28.57 | 28.57 |  0.00 | 57.14
                 | 50.00 | 50.00 |  0.00 |
                 |100.00 | 66.67 |  0.00 |
        ---------+--------+--------+--------+
        sa       |    0 |     1 |     2 |      3
                 |  0.00 | 14.29 | 28.57 | 42.86
                 |  0.00 | 33.33 | 66.67 |
                 |  0.00 | 33.33 |100.00 |
        ---------+--------+--------+--------+
        Total          2       3       2       7
                   28.57   42.86   28.57  100.00
```

(1)

```
                 Statistics for Table of treat by satisf

        Statistic                 DF      Value      Prob
        -------------------------------------------------------
        Chi-Square                 2     4.2778    0.1178
        Likelihood Ratio Chi-Square 2    5.7416    0.0567
        Mantel-Haenszel Chi-Square  1    3.5000    0.0614
        Phi Coefficient                  0.7817
        Contingency Coefficient          0.6159
        Cramer's V                       0.7817

     WARNING: 100% of the cells have expected counts less than 5.
              (Asymptotic) Chi-Square may not be a valid test.


                   Pearson Chi-Square Test
              ---------------------------------
              Chi-Square                 4.2778
              DF                              2
              Asymptotic Pr >  ChiSq     0.1178
              Exact      Pr >= ChiSq     0.3143


                Likelihood Ratio Chi-Square Test
              ---------------------------------
              Chi-Square                 5.7416
              DF                              2
              Asymptotic Pr >  ChiSq     0.0567
              Exact      Pr >= ChiSq     0.3143


                Mantel-Haenszel Chi-Square Test
              ---------------------------------
              Chi-Square                 3.5000
              DF                              1
              Asymptotic Pr >  ChiSq     0.0614
              Exact      Pr >= ChiSq     0.1714

                   Sample Size = 7
```

For larger tables, it may be desirable to obtain a random sample of permutations to reduce computation time. To do this, alter the

exact chisq;

statement as

exact chisq / mc;

which requests 10,000 random samples. Or, to request any number of samples, use

exact chisq / n = *number of samples*;

6.3   Fisher's Exact Test for a 2 × 2 Contingency Table

Fisher's exact test is a permutation test applied to a 2×2 contingency table. To illustrate the computations, consider Table 6.3.1. The statistic X computed by SAS is the number in Row 1, Column 1, although the number in any of the cells could be used. A known value of X determines all other entries in the table, given that the row and column totals are fixed. For instance, if $X = 3$, then the entry in Row 1, Column 2 is 8; the entry in Row 2, Column 1 is 4; and the entry in Row 2, Column 2 is 4.

Table 6.3.1. A 2×2 contingency table with fixed row and column totals.

	C1	C2	Total
R1	X		11
R2			8
Total	7	12	19

If we permute the 7 column labels C1 and the 12 column labels C2 among the rows, 11 to Row 1 and 8 to Row 2, then the probability that $X = k$ is given by the hypergeometric probability

$$P_F(X = k) = \frac{\binom{7}{k}\binom{12}{11-k}}{\binom{19}{11}}.$$

If X_{obs} is the observed value of X, then the lower-tail p-value is $P_F(X \le X_{obs})$. The inequality is reversed for the upper-tail p-value. For a two-sided p-value, SAS sums all probabilities less than or equal to $P_F(X = X_{obs})$.

SAS Implementation
When **PROC FREQ** is used to analyze 2×2 contingency tables, Fisher's exact test is automatically displayed.

Example 6.3.1. A surgeon is interested in determining whether the administration of a certain drug can reduce the incidence of pulmonary embolism in patients undergoing high-risk surgery. Nineteen patients were selected for the study, with 11 getting the drug and 8 receiving the standard treatment. The data are shown in Table 6.3.2. A favorable outcome for the drug would be a low incidence of pulmonary embolism, and thus a one-sided test is desirable.

Table 6.3.2. Incidence of pulmonary embolism (P.E.).			
	P.E.	No P.E.	Total
Drug	3	8	11
No Drug	4	4	8
Total	7	12	19

The following code produces Fisher's exact test for the data of Table 6.3.2.

```
data ta6_3_2;
input Drug PE count @@;
datalines;
1 1 3   1 2 8
2 1 4   2 2 4
;

proc freq data=ta6_3_2;
weight count;              /* Indicates frequency variable */
exact chisq;               /* Requests Fisher's exact test */
tables Drug*PE;
run;
```

The exact one-sided p-value corresponding to the probability of three or fewer in the Drug–P.E. category **(1)** is 0.2966 where the frequency in row 1, column 1 is used by SAS as the test statistic. If we were interested in the probability of three or more in the Drug–P.E. category, then the p-value **(2)** is 0.9326. The two-sided p-value **(3)** 0.3765 is equivalent to that given by the permutation chi-square test **(4)**. Finally, 0.2292 **(5)** is the probability of obtaining exactly 3 in Row 1, Column 1 of the table.

Output – Example 6.3.1:

```
                        The FREQ Procedure

                      Table of Drug by PE

              Drug        PE

              Frequency|
              Percent  |
              Row Pct  |
              Col Pct  |      1|      2|  Total

                     1 |     3 |     8 |    11
                       | 15.79 | 42.11 | 57.89
                       | 27.27 | 72.73 |
                       | 42.86 | 66.67 |

                     2 |     4 |     4 |     8
                       | 21.05 | 21.05 | 42.11
                       | 50.00 | 50.00 |
                       | 57.14 | 33.33 |

              Total         7      12      19
                        36.84   63.16  100.00
```

```
           Statistics for Table of Drug by PE

     Statistic                    DF     Value    Prob

     Chi-Square                    1    1.0281   0.3106
     Likelihood Ratio Chi-Square   1    1.0269   0.3109
     Continuity Adj. Chi-Square    1    0.2834   0.5945
     Mantel-Haenszel Chi-Square    1    0.9740   0.3237
     Phi Coefficient                   -0.2326
     Contingency Coefficient            0.2266
     Cramer's V                        -0.2326
```

```
     WARNING: 50% of the cells have expected counts less than 5.
              (Asymptotic) Chi-Square may not be a valid test.
```

```
                   Pearson Chi-Square Test

        Chi-Square                       1.0281
        DF                                    1
        Asymptotic Pr >  ChiSq           0.3106
        Exact        Pr >= ChiSq         0.3765
```

(4)

```
              Likelihood Ratio Chi-Square Test

        Chi-Square                       1.0269
        DF                                    1
        Asymptotic Pr >  ChiSq           0.3109
        Exact        Pr >= ChiSq         0.3765
```

Output – Example 6.3.1 (continued):

```
            Mantel-Haenszel Chi-Square Test
         ─────────────────────────────────────
         Chi-Square                      0.9740
         DF                                   1
         Asymptotic  Pr >  ChiSq        0.3237
         Exact       Pr >= ChiSq        0.3765


                 Fisher's Exact Test
         ─────────────────────────────────────
         Cell (1,1) Frequency (F)             3
         Left-sided Pr <= F              0.2966
         Right-sided Pr >= F             0.9326

         Table Probability (P)          0.2292
         Two-sided Pr <= P              0.3765

                 Sample Size = 19
```

(with line labels to the left: (1) for Left-sided Pr <= F, (2) for Right-sided Pr >= F, (5) for Table Probability (P), (3) for Two-sided Pr <= P)

6.4   Contingency Tables with Ordered Categories

If the categories of one or both of the factors in an $R \times C$ contingency table are ordered, then there are tests that may be more powerful than the Pearson or likelihood ratio chi-square tests for detecting associations among the row and column factors. If the column (or row) factor is ordered, but the other is not, then the table is *singly ordered*. Typically, such tables occur when the row factor represents a treatment or a grouping variable, and the response is ordinal such as poor, fair, good, or excellent. If both factors are ordered, then the table is *doubly ordered*. Such tables typically occur when one factor is quantitative, such as levels of doses of a drug, and the other factor is ordinal, such as the level of pain relief produced by the drug.

6.4.1 Singly Ordered Tables

Example 6.4.1. Department heads of three colleges were asked to rate the performance of the chief academic officer. The rating scale used in the study was poor, fair, good, very good, and excellent. The results are displayed as a contingency table in Table 6.4.1.

Table 6.4.1. Ratings of the Chief Academic Officer.				
	Poor (1)	Fair (2)	Good (3)	Very Good (4)	Excellent (5)
College A	6	4	3	2	0
College B	2	2	3	2	1
College C	1	2	3	4	2

The responses for College A tend to be "smaller" in the sense that the majority are in the poor-to-fair range, whereas the responses for Colleges B and C tend to be "larger" in the sense that the majority are in the good-to-excellent range. The Kruskal-Wallis test may be used to detect differences of this type.

SAS Implementation

There are two procedures in SAS that can be used to produce the Kruskal-Wallis test: **PROC NPAR1WAY** and **PROC FREQ**. The advantage of using **PROC NPAR1WAY** is that an exact *p*-value is available. However, raw data are required, and if the data are in contingency table format, then the data first have to be converted to raw form. Section 3.2 describes implementation of the Kruskal-Wallis test in **PROC NPAR1WAY**. **PROC FREQ**, however, accepts either raw data or data in contingency table format, but will only produce an asymptotic *p*-value. The test is based on the generalized Cochran-Mantel-Haenszel (CMH) statistic.

The following code uses **PROC FREQ** to produce the Kruskal-Wallis test for the data of Table 6.4.1. The **scores=ranks** option requests analysis on ranks, which produces the Kruskal-Wallis test (the default is to use the raw data and thus produce a one-way ANOVA *F*-test, treating the response as interval level). Although the values given for the ordered category do not need to be numeric, the coding must have the same ordering as the categories. For example, if we had used the string values "P," "F," "G," "VG," and "E" to represent the categories, then SAS would order these alphabetically as E, F, G, P and VG, and the resulting analysis would not be correct. Because the data are given in a contingency table, this method of input will be used.

```
data ta6_4_1;
input college $ rating count @@;
datalines;
A 1 6 A 2 4 A 3 3 A 4 2 A 5 0
B 1 2 B 2 2 B 3 3 B 4 2 B 5 1
C 1 1 C 2 2 C 3 3 C 4 4 C 5 2
;

proc freq data=ta6_4_1;
tables college*rating /
          chisq                   /* Requests chi-square tests */
          cmh                     /* Requests Cochran-Mantel-Haenszel tests */
          scores=ranks            /* Requests analysis on ranks */
          noprint;                /* Suppresses printing of tables */
weight count;                     /* Indicates frequency variable */
run;
```

The Pearson chi-square statistic **(1)** is $\chi^2 = 7.02$ with exact p-value 0.534, while the Kruskal-Wallis test **(2)**, given in the second row of the Cochran-Mantel-Haenszel section, is $KW = CMH_2 = 6.43$ with *p*-value 0.04, providing much stronger evidence of rating differences among colleges than given by the Pearson chi-square statistic. The third row of the CMH section also gives a test of no association **(3)** that is asymptotically equivalent to the chi-square test. The test statistic is

$$CMH_3 = \left(\frac{n-1}{n}\right)\chi^2 = \left(\frac{36}{37}\right)7.02 = 6.833 .$$

The test for nonzero correlation in the first row of the CMH section is the same as the test based on the Mantel-Haenszel chi-square statistic. It is not appropriate for singly ordered tables.

Output –Example 6.4.1:

```
                          The FREQ Procedure

                 Statistics for Table of college by rating

              Statistic                   DF     Value     Prob

(1)           Chi-Square                   8    7.0231    0.5341
              Likelihood Ratio Chi-Square  8    8.1354    0.4204
              MH Chi-Square (Rank Scores)   1    6.4138    0.0113
              Phi Coefficient                   0.4357
              Contingency Coefficient           0.3994
              Cramer's V                        0.3081


              WARNING: 100% of the cells have expected counts less
                       than 5. Chi-Square may not be a valid test.

                          Sample Size = 37


                 Summary Statistics for college by rating

          Cochran-Mantel-Haenszel Statistics (Based on Rank Scores)

          Statistic   Alternative Hypothesis    DF    Value    Prob

              1       Nonzero Correlation        1    6.4138   0.0113
(2)           2       Row Mean Scores Differ     2    6.4283   0.0402
(3)           3       General Association        8    6.8333   0.5547
```

An alternative way to obtain the Kruskal-Wallis test is to convert the tabled data to raw scores and apply **PROC NPAR1WAY**. The following code and output show this for the data in Table 6.4.1. The Monte Carlo estimate of the exact *p*-value is 0.0498 **(1)**. Note that the asymptotic *p*-value of 0.0402 **(2)** is the same as given for the "Row Mean Scores Differ" test from the previous output.

```
/* New data set format for KW test */
data ta6_4_1raw;
input college $ resp @@;
datalines;
A 1 A 1 A 1 A 1 A 1 A 1 A 2 A 2 A 2 A 2 A 3 A 3 A 3 A 3 A 4 A 4
B 1 B 1 B 2 B 2 B 3 B 3 B 3 B 3 B 4 B 4 B 5
C 1 C 2 C 2 C 3 C 3 C 3 C 4 C 4 C 4 C 4 C 5 C 5
;

proc npar1way data=ta6_4_1raw wilcoxon;
class college;
exact wilcoxon/              /* Requests exact test on ranks */
      mc;                    /* Requests 10,000 random permutations */
var resp;
run;
```

Output – Example 6.4.1 (continued):

The NPAR1WAY Procedure

Wilcoxon Scores (Rank Sums) for Variable resp
Classified by Variable college

college	N	Sum of Scores	Expected Under H0	Std Dev Under H0	Mean Score
A	15	211.0	285.0	31.527194	14.066667
B	10	200.0	190.0	28.517420	20.000000
C	12	292.0	228.0	30.060000	24.333333

Average scores were used for ties.

Kruskal-Wallis Test

Chi-Square	6.4283
DF	2
(2) Pr > Chi-Square	0.0402

Monte Carlo Estimate for the Exact Test

Pr >= Chi-Square	
(1) Estimate	0.0408
99% Lower Conf Limit	0.0357
99% Upper Conf Limit	0.0459
Number of Samples	10000
Initial Seed	61254

6.4.2 Cochran-Armitage Test

The Cochran-Armitage test is a test for trend when data are arranged in an $R \times 2$ or $2 \times C$ table. For illustration, assume that the data are in an $R \times 2$ table. The rows are assumed to have numerical labels $L_1, L_2, ..., L_R$ which, for instance, could represent doses of a drug. The columns are assumed to be the Bernoulli responses "success" and "failure," where the successes are in the first column. Suppose that the data for ith individual, $i = 1, ..., n$, are denoted as (X_i, Y_i), where X_i is the numerical label of the row in which the ith individual is classified, and Y_i is either 1 for "success" or 0 for "failure." Let b_1 denote the slope of the least squares regression line for the n pairs (X_i, Y_i). The Cochran-Armitage statistic is $b_1 / \sqrt{Var(b_1)}$, where $Var(b_1)$ is the estimated variance of the slope estimate under the null hypothesis of no trend. The variance estimate is

$$Var(b_1) = \hat{p}(1-\hat{p}) \bigg/ \sum_{i=1}^{n} (X_i - \overline{X})^2 ,$$ where $\hat{p}$ is fraction of successes among all rows.

SAS Implementation

The Cochran-Armitage test for trend can be implemented using **PROC FREQ**.

Example 6.4.2. Suppose that in Example 6.3.1, instead of simply "Drug" or "No Drug," four levels of drug dose were used: No Drug (0), Low Dose (1), Medium Dose (2), and High Dose (3). The results for a random sample of 39 patients are given in Table 6.4.2. A favorable outcome for the drug would be a low incidence of pulmonary embolism, and thus a one-sided test is desirable.

Table 6.4.2. Incidence of pulmonary embolism (P.E.).			
	P.E.	No P.E.	Total
High dose	1	9	10
Medium dose	3	8	11
Low dose	4	6	10
No drug	4	4	8
Total	12	27	39

The following code requests the Cochran-Armitage test for the data of Table 6.4.2. The **trend** option in the **tables** statement requests the CA test.

```
data ta6_4_2;
input dose resp $ count @@;
datalines;
3 PE 1 3 NPE 9
2 PE 3 2 NPE 8
1 PE 4 1 NPE 6
0 PE 4 0 NPE 4
;

proc freq data=ta6_4_2;
weight count;              /* Indicates frequency variable */
tables dose*resp /
      trend               /* trend option requests CA test */
      noprint;            /* Suppresses printing of tables */
exact trend;              /* Requests exact test */
run;
```

The Cochran-Armitage statistic is 1.9533 with exact one-sided *p*-value 0.0382.

Output –Example 6.4.2:

```
                        The FREQ Procedure

                Statistics for Table of dose by resp

                    Cochran-Armitage Trend Test
                    ---------------------------
                    Statistic (Z)        1.9533

                    Asymptotic Test
                    One-sided Pr >  Z    0.0254
                    Two-sided Pr > |Z|   0.0508

                    Exact Test
                    One-sided Pr >= Z    0.0382
                    Two-sided Pr >= |Z|  0.0575

                      Sample Size = 39
```

6.4.3 Doubly Ordered Tables

The Jonckheere-Terpstra test discussed in Section 3.4 and the Mantel-Haenszel chi-square statistic discussed in Section 6.1 are appropriate for doubly ordered tables. Two versions of the Mantel-Haenszel statistic may be considered. One is appropriate when the columns labels are numerical. The statistic is $(n-1)r^2$, where r is the Pearson product-moment correlation. If a factor is ordinal but non-numeric, then SAS assigns the numerical value 1 for the category label with the lowest alpha-numeric ordering, 2 to the category label with the next lowest alpha-numeric ordering, and so forth in computing the correlation. Another version of the Mantel-Haenszel statistic, which is appropriate for both numerical and ordinal categories, is $(n-1)r_S^2$, where r_S is the Spearman correlation. In the computation of the statistic, the categories are ranked according to their alpha-numeric ordering, and average ranks are assigned to the ties.

SAS Implementation

The Jonckheere-Terpstra and Mantel-Haenszel tests can be generated in **PROC FREQ**.

Example 6.4.3. An entomologist was interested in comparing the effectiveness of three methods of controlling insect damage to alfalfa plants. Plants were infected with insects and treated with product A, B, or C. The damage done by the insects to each plant was classified as severe, moderate, slight, or none. It was anticipated that the plants with treatment A would have the most damage, followed by those with treatment B, and then C. The data are shown in Table 6.4.3.

Table 6.4.3. Level of insect damage for each of three treatments.

Response/Treatment	Severe	Moderate	Slight	None
A	10	12	17	30
B	9	9	11	35
C	7	8	12	43

The following code generates the Pearson chi-square statistic, the likelihood ratio chi-square statistic, the Mantel-Haenszel chi-square statistic based on ranks, the JT statistic, and the Kruskal-Wallis statistic for the data in Table 6.4.3.

```
data ta6_4_3;
input treat resp count @@;
datalines;
1 1 10 1 2 12 1 3 17 1 4 30
2 1 9 2 2 9 2 3 11 2 4 35
3 1 7 3 2 8 3 3 12 3 4 43
;

proc freq data=ta6_4_3;
weight count;
tables treat*resp /
        chisq              /* Requests chi-square tests */
        jt                 /* Requests Jonckheere-Terpstra test */
        cmh                /* Requests Cochran-Mantel-Haenszel tests */
        scores=ranks       /* Requests analysis on ranks */
        noprint;
exact chisq jt /           /* Requests exact permutation p-values */
        mc;                /* Requests 10,000 random permutations */
run;
```

The two-sided Monte Carlo p-values for the Mantel-Haenszel chi-square test based on ranks **(2)** and the JT test **(3)** are 0.0508 and 0.0477, respectively. These estimates are based on 10,000 randomly selected permutations. An advantage of the JT test is that a one-sided p-value is also available (0.0243). The effect of using tests that account for the doubly ordered property of the table can be seen by comparing these p-values to those of the Pearson chi-square test **(1),** which has a Monte Carlo p-value 0.5519, and the Kruskal-Wallis test **(5),** which has an asymptotic p-value 0.1382. The Mantel-Haenszel chi-square test based on ranks is computed when the **scores=ranks** option is invoked, and it is computed under both the **chisq** and **cmh** options. In the output for the **cmh** option, this test is labeled as "Nonzero Correlation" **(4).**

Output –Example 6.4.3:

The FREQ Procedure

Statistics for Table of treat by resp

Statistic	DF	Value	Prob
Chi-Square	6	4.9649	0.5483
Likelihood Ratio Chi-Square	6	4.9872	0.5455
MH Chi-Square (Rank Scores)	1	3.9516	0.0468
Phi Coefficient		0.1564	
Contingency Coefficient		0.1545	
Cramer's V		0.1106	

Output –Example 6.4.3 (continued):

```
                    Pearson Chi-Square Test
        ────────────────────────────────────────────
        Chi-Square                      4.9649
        DF                                   6
        Asymptotic Pr >  ChiSq          0.5483
```

Monte Carlo Estimate for the Exact Test

(1)
```
        Pr >= ChiSq                     0.5585
        99% Lower Conf Limit            0.5457
        99% Upper Conf Limit            0.5713

        Number of Samples               10000
        Initial Seed                    76817
```

```
                Likelihood Ratio Chi-Square Test
        ────────────────────────────────────────────
        Chi-Square                      4.9872
        DF                                   6
        Asymptotic Pr >  ChiSq          0.5455
```

Monte Carlo Estimate for the Exact Test

```
        Pr >= ChiSq                     0.5567
        99% Lower Conf Limit            0.5439
        99% Upper Conf Limit            0.5695

        Number of Samples               10000
        Initial Seed               1607406312
```

```
            Statistics for Table of treat by resp

                Mantel-Haenszel Chi-Square Test
                        (Rank Scores)
        ────────────────────────────────────────────
        Chi-Square                      3.9516
        DF                                   1
        Asymptotic Pr >  ChiSq          0.0468
```

Monte Carlo Estimate for the Exact Test

(2)
```
        Pr >= ChiSq                     0.0508
        99% Lower Conf Limit            0.0451
        99% Upper Conf Limit            0.0565

        Number of Samples               10000
        Initial Seed                557917355
```

Output –Example 6.4.3 (continued)**:**

```
                         Jonckheere-Terpstra Test
                  ─────────────────────────────────────
                  Statistic (JT)       7686.5000
                  Z                       1.9763
                  One-sided Pr >  Z       0.0241
                  Two-sided Pr > |Z|      0.0481

                  Monte Carlo Estimates for the Exact Test
```

(3)
```
                  One-sided Pr >=  JT
                  Estimate                0.0243
                  99% Lower Conf Limit    0.0203
                  99% Upper Conf Limit    0.0283


                  Two-sided Pr >= |JT|
                  Estimate                0.0477
                  99% Lower Conf Limit    0.0422
                  99% Upper Conf Limit    0.0532


                  Number of Samples         10000
                  Initial Seed         2017428669


                       Sample Size = 203



                  Summary Statistics for treat by resp

          Cochran-Mantel-Haenszel Statistics (Based on Rank Scores)


          Statistic   Alternative Hypothesis    DF    Value    Prob
          ───────────────────────────────────────────────────────────
```
(4)
(5)
```
              1       Nonzero Correlation        1    3.9516   0.0468
              2       Row Mean Scores Differ     2    3.9586   0.1382
              3       General Association        6    4.9405   0.5515

                       Total Sample Size = 203
```

6.5   Tables with Multiple Strata

Let S_k, $k = 1,...,s$, denote a set of s $R \times C$ contingency tables in which the row and column factors are the same for all tables. The different tables represent *strata*. The null hypothesis of interest is that there is no relationship between the row and column factors in all the tables. The alternatives of interest are ones in which the associations between the row and column factors are similar across the strata.

Example 6.5.1. A study was done at three different centers to compare two procedures, I and II, for treating a particular medical condition. Each patient's outcome was classified as either "complications occur" (C) or "no complications occur" (NC). Data are summarized in Table 6.5.1, which shows three strata, one for each center. It is expected that the complication rate will be lower for procedure I than for procedure II, but the extent to which it is lower may depend on the center.

Table 6.5.1. Comparison of the effect of procedures I and II on the incidence of complications (C = complications, NC = no complications). Entries are the number in each category.

Center 1			
Procedure	C	NC	Row Totals
I	2	8	10
II	7	8	15
Column Totals	9	16	25
Center 2			
Procedure	C	NC	Row Totals
I	3	9	12
II	14	28	42
Column Totals	17	37	54
Center 3			
Procedure	C	NC	Row Totals
I	1	9	10
II	13	17	30
Column Totals	14	26	40

Stratified contingency tables may be analyzed using **PROC FREQ** with the **cmh** option. The statistic that is computed under the **cmh** option is a quadratic form that involves the following: (1) the $1 \times RC$ vectors of the form $(n_{11k}, \ldots, n_{RCk})$, $k = 1, \ldots, s$, (2) the estimated mean and covariance matrices of these vectors under the null hypothesis of no association between row and column factors, and (3) the scores attached to the rows and columns. In the case of 2×2 tables, the CMH

statistic can be expressed as $Z^2 = \left(\sum_{k=1}^{s} (n_{11k} - E(n_{11k})) \right)^2 / \sum_{k=1}^{s} Var(n_{11k})$, where n_{11k} is the count in

cell $(1,1)$ of the kth strata, and the expected values and variances are those of the hypergeometric distributions of the n_{ijk}'s.

The asymptotic distribution of the CMH statistic is chi-square, and in the case of stratified 2×2 tables, there is one degree of freedom. A permutation test for the CMH statistic is not available under the **cmh** option, although permutation tests may be done separately for each table by including the **chisq** with the **tables** statement and including the **exact chisq** statement.

Suppose a 2×2 table consists of treatment and control categories for the rows and success and failure categories for the columns. Let P_T denote the probability of success for the treatment and P_C denote the probability of success for the control. Then the *odds ratio* is $P_T(1 - P_C)/[P_C(1 - P_T)]$. The ratio of failure probabilities is $(1 - P_T)/(1 - P_C)$ and is called the *relative risk of failure*. The ratio of success probabilities is P_T / P_C and is called the *relative risk of success*. SAS provides the Mantel-Haenszel and logit estimates of these quantities assuming that they are the same across strata. SAS also provides the Breslow-Day test for homogeneity of odd-ratios across strata.

For a discussion of the CMH statistic, odds ratios, and relative risks, see Agresti (2002).

SAS Implementation
The Cochran-Mantel-Haenszel test can be implemented in **PROC FREQ**. The **cmh** option in the **tables** statement is used to produce the test.

The following code requests the Cochran-Mantel-Haenszel test on data of Table 6.5.1. SAS recognizes the first variable specified in the **tables** statement as the stratum variable. Thus, in this example, "center" is listed first.

```
data ta6_5_1;
input center proc outcome $ count @@;
datalines;
1 1 c 2
1 1 nc 8
1 2 c 7
1 2 nc 8
2 1 c 3
2 1 nc 9
2 2 c 14
2 2 nc 28
3 1 c 1
3 1 nc 9
3 2 c 13
3 2 nc 17
;

proc freq data=ta6_5_1;
weight count;
tables center*proc*outcome/        /* "center" listed first to denote strata */
       cmh                         /* Requests Cochran-Mantel-Haenszel tests */
       noprint;                    /* Suppresses printing of tables */
run;
```

The CMH statistic **(1)** is 4.5395 with *p*-value 0.0331. SAS gives only the *p*-value based on the chi-square approximation, not the exact *p*-value. Also displayed are estimates and confidence intervals of the common odds ratio and relative risks **(2)**.

Output –Example 6.5.1:

```
                        The FREQ Procedure

                 Summary Statistics for proc by outcome
                        Controlling for center

           Cochran-Mantel-Haenszel Statistics (Based on Table Scores)

           Statistic   Alternative Hypothesis    DF    Value    Prob
           ────────────────────────────────────────────────────────────
               1        Nonzero Correlation       1    4.5395   0.0331
               2        Row Mean Scores Differ     1    4.5395   0.0331
(1)            3        General Association        1    4.5395   0.0331
```

Output –Example 6.4.3 (continued):

(2)   Estimates of the Common Relative Risk (Row1/Row2)

Type of Study	Method	Value	95% Confidence Limits	
Case-Control	Mantel-Haenszel	0.3495	0.1297	0.9415
(Odds Ratio)	Logit	0.3728	0.1352	1.0279
Cohort	Mantel-Haenszel	0.4676	0.2164	1.0101
(Col1 Risk)	Logit	0.5171	0.2400	1.1141
Cohort	Mantel-Haenszel	1.3568	1.0588	1.7385
(Col2 Risk)	Logit	1.3734	1.0761	1.7530

Breslow-Day Test for
Homogeneity of the Odds Ratios

Chi-Square	1.4698
DF	2
Pr > ChiSq	0.4795

Total Sample Size = 119

6.6   McNemar's Test

McNemar's test is appropriate for paired data where the responses are dichotomous. We may consider each pair to consist of responses of the form (X_i, Y_i), where (X_i, Y_i) is one of the following 4 possibilities: (0,0), (0,1), (1,0), and (1,1). If $p_1 = P(X_i = 1)$ and $p_2 = P(Y_i = 1)$, then the hypotheses of interest are $H_0 : p_1 = p_2$ versus one-sided or two-sided alternatives. We may summarize the results in a 2×2 contingency table. The categories for the row factor are the 2 possible responses for the X's, and the categories for the column factor are the 2 possible responses for the Y's. Let n_{ij} denote the entry in the (ij)th cell of the 2×2 table. A test statistic for testing either a one-sided or two-sided hypotheses is $Z = \dfrac{n_{12} - n_{21}}{\sqrt{n_{12} + n_{21}}}$, which has an asymptotic standard normal distribution. SAS computes the square of this statistic.

SAS Implementation
The **FREQ** procedure will produce McNemar's test.

Example 6.6.1. Table 6.6.1 summarizes responses for 20 chemotherapy patients who were asked whether or not they experienced nausea before and after being given a drug that is intended to reduce nausea. In the table below, there are 12 out of 20 who experienced nausea before being given the drug and 5 out of 20 afterward, indicating that the drug may be effective in reducing the incidence of nausea.

Table 6.6.1. Two-way table for paired data.		
	Nausea After	No Nausea After
Nausea before	3	9
No nausea before	2	6

The following code requests McNemar's test for the data of Table 6.6.1. The **mcnem** option in the **exact** statement requests both asymptotic and exact *p*-values.

```
data ta6_6_1;
input before $ after $ count @@;
datalines;
A A 3 A B 9
B A 2 B B 6
;
proc freq data=ta6_6_1;
weight count;
exact mcnem;                          /* Requests McNemar test, exact p-value */
tables before*after / noprint;
run;
```

McNemar's statistic is 4.4545 with corresponding exact *p*-value 0.0654.

Output –Example 6.6.1:

```
                    The FREQ Procedure

            Statistics for Table of before by after

                        McNemar's Test
            ──────────────────────────────────────
            Statistic (S)         4.4545
            DF                         1
            Asymptotic Pr >  S    0.0348
            Exact       Pr >= S   0.0654


                 Simple Kappa Coefficient
            ──────────────────────────────────────
            Kappa                 -0.0000
            ASE                    0.1725
            95% Lower Conf Limit  -0.3381
            95% Upper Conf Limit   0.3381

                 Sample Size = 20
```

7

Analysis of Censored Data

Suppose that T denotes the lifetime of an individual or the time to failure of an object such as a lightbulb or battery. An observation is *right-censored* at time t_0 if $T > t_0$ but the value of T is unknown. For instance, let T denote the survival time of a patient after undergoing chemotherapy. If the patient decides to discontinue treatment after 10 months and there is no follow-up contact with the patient, then the patient's survival time is censored at 10 months. An observation censored at time t_0 is denoted by the plus superscript to the right of the observation, that is, t_0^+. It is also possible to have *left-censoring*. For instance, an instrument designed to detect pollutants may register that a pollutant is present, but the amount is so small that the exact value is only known to be less than a threshold reading t_L. We will only discuss right-censored data.

Different modes of right-censoring may occur. In studies on patients undergoing medical treatments, censoring is often random; that is, for reasons unrelated to the treatment, the patient may leave the study with no follow-up. In dealing with tests of mechanical or electrical components, censoring is usually either *Type I* or *Type II*. In Type I censoring, objects are tested for a fixed period of time, say 24 hours. All items that are still working after that time have censored survival times. In Type II censoring, n items are placed on test, and testing stops after the first k have failed.

7.1   Estimating the Survival Function

Let $F(t)$ denote the cumulative distribution function of a lifetime random variable T. The survival function is $S(t) = P(T > t) = 1 - F(t)$. The *Kaplan-Meier* or *product-limit estimate* of the survival function when data are censored is implemented in SAS. The procedure is defined iteratively as follows. Let $t_{(1)} \le t_{(2)} \le ... \le t_{(n)}$ denote the numerical values, including ties and censored observations, placed in order from smallest to largest where the censoring notation is suppressed. Assume that $t_{(1)}$ is an uncensored time and let

$$\hat{S}\left(t_{(1)}\right) = \textit{fraction of observations} > t_1.$$

If $t_{(i)} < t_{(j)}$ are adjacent uncensored times, $i > 1$, then define

$$\hat{S}\left(t_{(j)}\right) = \hat{S}\left(t_{(i)}\right) \frac{\textit{fraction of observations} > t_{(j)}}{\textit{fraction of observations} \ge t_{(j)}}$$

For a time t between two adjacent uncensored observations $t_{(i)} < t_{(j)}$, we let $\hat{S}(t) = \hat{S}\left(t_{(i)}\right)$. See Lee (2003) for more details.

SAS Implementation

The **LIFETEST** procedure can be used to estimate the survival function when data are right-censored.

Example 7.1.1. Table 7.1.1 gives survival times for each of nine patients.

Table 7.1.1. Survival times of nine patients. The plus ("+") to the right of the time indicates a censored observation.								
Patient	1	2	3	4	5	6	7	8	9
Time	3.1	4.0^+	5.3	5.3	6.1^+	7.5	7.5^+	8.9	9.0

The following code generates the Kaplan-Meier estimates of the survival function for the data of Table 7.1.1. The **time** statement is required and gives the survival (or "failure") time variable, in this case called "time." Following the asterisk is a censoring variable, and values listed in parentheses following the censoring variable indicate censored times.

```
data ta7_1_1;
input time status @@;
datalines;
3.1 1 4.0 0 5.3 1 5.3 1 6.1 0 7.5 1 7.5 0 8.9 1 9.0 1
;

proc lifetest data=ta7_1_1;
time time*status(0);        /* "Status=0" indicates censored observation */
run;
```

The estimated survival function values for the uncensored times are given in the column labeled "Survival."

Output – Example 7.1.1:

```
                        The LIFETEST Procedure


                      Product-Limit Survival Estimates

                                     Survival
                                     Standard    Number     Number
         time     Survival  Failure   Error      Failed      Left

       0.00000     1.0000      0         0          0          9
       3.10000     0.8889    0.1111    0.1048       1          8
       4.00000*       .         .         .         1          7
       5.30000        .         .         .         2          6
       5.30000     0.6349    0.3651    0.1692       3          5
       6.10000*       .         .         .         3          4
       7.50000     0.4762    0.5238    0.1871       4          3
       7.50000*       .         .         .         4          2
       8.90000     0.2381    0.7619    0.1926       5          1
       9.00000        0      1.0000      0          6          0

      NOTE: The marked survival times are censored observations.
```

Output – Example 7.1.1 (continued):

```
              Summary Statistics for Time Variable time

                      Quartile Estimates

                    Point      95% Confidence Interval
        Percent   Estimate      [Lower        Upper)

             75    8.90000      7.50000      9.00000
             50    7.50000      5.30000      9.00000
             25    5.30000      3.10000      8.90000


                    Mean     Standard Error

                  7.14286        0.78688
```

7.2   Permutation Tests for Type I and Type II Censored Data

Suppose that we have Type I or Type II censored data in which the largest and only the largest of the combined observations from two or more groups are censored. We may perform rank tests on the data by assigning tied ranks to the right-censored values. The data may then be analyzed using the Wilcoxon rank-sum test with ties if there are two groups or the Kruskal-Wallis test with ties if there are more than two groups.

SAS Implementation

The **NPAR1WAY** procedure can be used to produce the Wilcoxon rank-sum test for Type I or Type II censored data.

Example 7.2.1. Table 7.2.1 gives lifetimes (in hours) of two brands of batteries with $T = 4$ hours as censoring time. A plus ("+") to the right of the time indicates a censored observation. In this case, all observations at 4.0 are censored and will automatically be assigned the largest rank, adjusted for ties. If one value at 4.0 were not censored and the others were, then we would have to enter a value greater than 4.0, say 4.01, for the values that are censored at 4.0 to get the correct rankings.

Table 7.2.1. Lifetimes (in hours) of two brands of batteries. The plus ("+") to the right of the time indicates a censored observation.

Brand A	1.5	2.3	2.4	2.6	3.0	3.0	3.4	3.8
Brand B	2.5	2.9	3.3	3.7	3.9	4.0^+	4.0^+	4.0^+

The following code computes the Wilcoxon rank-sum test for the data of Table 7.2.1.

```
data ta7_2_1;
input brand $ lifetime@@;
datalines;
A 1.5 A 2.3 A 2.4 A 2.6 A 3.0 A 3.0 A 3.4 A 3.8
B 2.5 B 2.9 B 3.3 B 3.7 B 3.9 B 4.0 B 4.0 B 4.0
;
```

```
proc npar1way data=ta7_2_1
       wilcoxon;              /* Requests analysis on ranks */
class brand;
exact wilcoxon;               /* Requests exact p-value on ranks */
var lifetime;
run;
```

The exact two-sided *p*-value for the Wilcoxon rank-sum test is 0.0343.

***Output – Example 7.2.1*:**

```
                        The NPAR1WAY Procedure

            Wilcoxon Scores (Rank Sums) for Variable lifetime
                      Classified by Variable brand

                       Sum of      Expected       Std Dev         Mean
       brand    N      Scores      Under H0       Under H0        Score
      ─────────────────────────────────────────────────────────────────
       A        8       48.0         68.0        9.486833          6.0
       B        8       88.0         68.0        9.486833         11.0

                Average scores were used for ties.


                      Wilcoxon Two-Sample Test

              Statistic (S)                    48.0000

              Normal Approximation
              Z                                -2.0555
              One-Sided Pr <  Z                 0.0199
              Two-Sided Pr > |Z|                0.0398

              t Approximation
              One-Sided Pr <  Z                 0.0288
              Two-Sided Pr > |Z|                0.0577

              Exact Test
              One-Sided Pr <=  S                0.0172
              Two-Sided Pr >= |S - Mean|        0.0343
```

7.3   Tests for Randomly Censored Data

The general method for comparing groups with randomly censored data is to perform permutation tests on scores that have been assigned to the censored and uncensored observations. There are various methods of scoring. They all share the property that if the data are not censored, then the scores turn out to be equivalent to those produced by one of the standard methods of scoring discussed in Section 2.4. The two scoring systems most commonly generalized are the rank scores and the exponential or Savage scores. See Higgins (2004) for a discussion of scoring systems for randomly censored data.

We illustrate the method using *Gehan* scores, which is one of the first methods proposed for generalizing Wilcoxon scores to censored data. To compute a Gehan score of an observation, we compute the number observations definitely less than the observation in question minus the number of observations definitely greater than this observation. For instance, if the observations are 20, 30^+, 40, and 50, then the Gehan scores are -3, 1, 0, and 2. Once the scores are calculated, a permutation test may be applied to compare groups.

SAS Implementation

The **NPAR1WAY** procedure can be used to test for differences among survival functions. SAS does not have a routine for generating scores for censored data in the **NPAR1WAY** procedure, so they must be computed outside this procedure then entered into SAS.

Example 7.3.1. Table 7.3.1 gives survival times after diagnosis (in days) of two groups of patients, one undergoing a standard treatment (S) and the other a new treatment (N). The Gehan scores are indicated in parentheses.

Table 7.3.1. Survival times in days of two groups of patients. The plus ("+") to the right of the time indicates a censored observation. Gehan scores are in parentheses.

Standard treatment	94	155	180^+	375	741	951^+	1133	1198	1261
	(−16)	(−14)	(3)	(−9)	(−1)	(7)	(4)	(6)	(9)
New treatment	175	382	521	567^+	683^+	988	1216^+	1355^+	
	(−12)	(−7)	(−5)	(6)	(6)	(2)	(10)	(11)	

The following code generates a permutation test for testing the difference between the two treatments using the data of Table 7.3.1.

```
data ta7_3_1;
input treat $ time gehan @@;
datalines;
S 94 -16 S 155 -14 S 180 3  S 375 -9 S 741 -1
S 951 7  S 1133 4  S 1198 6 S 1261 9
N 175 -12 N 382 -7 N 521 -5 N 567 6
N 683 6 N 988 2 N 1216 10 N 1355 11
;

proc npar1way data=ta7_3_1
        scores=data;           /* Requests analysis on raw data */
class treat;
exact scores=data;             /* Requests exact p-value on raw data */
var gehan;
run;
```

The exact *p*-value for the test using Gehan scores is 0.2877.

Output – Example 7.3.1:

```
                       The NPAR1WAY Procedure

                   Data Scores for Variable gehan
                   Classified by Variable treat

                 Sum of      Expected      Std Dev         Mean
   treat    N    Scores      Under H0      Under H0        Score
   ─────────────────────────────────────────────────────────────
   S        9    -11.0         0.0       18.117265     -1.222222
   N        8     11.0         0.0       18.117265      1.375000


                 Data Scores Two-Sample Test

            Statistic (S)              11.0000

            Normal Approximation
            Z                           0.6072
            One-Sided Pr >  Z           0.2719
            Two-Sided Pr > |Z|          0.5437

            Exact Test
            One-Sided Pr >=  S          0.2877
            Two-Sided Pr >= |S - Mean|  0.5729
```

7.4 Comparing Survival Functions

7.4.1 Generalized Wilcoxon and Log-Rank Tests

These tests are based on a contingency table of group by censored status at each survival time as shown below for the two-group case:

Table 7.4.1. Comparison of number of survivors for two groups at survival time $t_{(i)}$.

Survived/Group	Group 1	Group 0	Total
No	d_{1i}	d_{0i}	d_i
Yes	$n_{1i} - d_{1i}$	$n_{0i} - d_{0i}$	$n_i - d_i$
Total	n_{1i}	n_{0i}	n_i

The expected number in a particular cell is calculated in the usual way as $\hat{e}_{ij} = \dfrac{n_{ij}d_i}{n_i}$, and variance

of d_{ij} is given by $\hat{v}_{jl} = \dfrac{n_{1i}n_{0i}d_i(n_i - d_i)}{n_i^2(n_i - 1)}$. Then the large sample test statistic can be defined as

$$\frac{\left[\sum_{i=1}^{m} w_i (d_{1i} - \hat{e}_{1i})\right]^2}{\sum_{i=1}^{m} w_i^2 \hat{v}_{1i}}.$$

This statistic can be interpreted as a weighted sum, taken over all survival times, of observed minus expected numbers of failure under the null hypothesis that the survival curves are identical, where the weights are $w_i = n_i$ for the Wilcoxon test and $w_i = 1$ for the log-rank test. For large samples, this statistic has an approximate chi-square distribution with 1 degree of freedom.

7.4.2 Likelihood Ratio Test

This statistic is based on the assumption that the observations in the various groups have an exponential distribution and tests for equality of the scale parameter across groups. The statistic that SAS uses is

$$2N \log\left(\frac{T}{N}\right) - 2\sum_j N_j \log\left(\frac{T_j}{N_j}\right),$$

where N_j is the total number of events in the jth group, N is the total number of events summed over all groups, T_j is the total time on test in the jth group, and $T = \sum_j T_i$.

See Lawless (2002) for more details about tests for samples from exponentially distributed populations.

SAS Implementation

The **LIFETEST** procedure generates the generalized Wilcoxon, log-rank, and likelihood ratio tests for comparing survival functions. The p-values given are asymptotic approximations.

Example 7.4.1. Table 7.3.1 gave survival times (in days) of two groups of patients.
The following code generates generalized Wilcoxon, log-rank, and likelihood ratio tests using the data of Table 7.3.1.

```
data ta7_3_1;
input treatment $ days status @@;
datalines;
S 94 1 S 155 1 S 180 0 S 375 1 S 741 1 S 951 0 S 1133 1 S 1198 1 S 1261 1
N 175 1 N 382 1 N 521 1 N 567 0 N 683 0 N 988 1 N 1216 0 N 1355 0
;

proc lifetest data=ta7_3_1;
time days*status(0);          /* Status=0 indicates censored observation */
strata treatment;            /* Identifies class variable for comparisons */
run;
```

The two-sided p-values are given in the partial output below. The p-value using the Wilcoxon scores is 0.5472, and using the log-rank scores is 0.3798. The p-value for the likelihood ratio test is 0.3926.

Output – Example 7.4.1:

Test of Equality over Strata

Test	Chi-Square	DF	Pr > Chi-Square
Log-Rank	0.7712	1	0.3798
Wilcoxon	0.3623	1	0.5472
-2Log(LR)	0.7310	1	0.3926

The test statistics above can be generalized to more than two groups. The **LIFETEST** procedure also generates the generalized Wilcoxon, log-rank, and likelihood ratio tests for comparing more than two survival functions. For these cases, the degrees of freedom for the chi-square distribution is *number of groups* – 1.

Example 7.4.2. Table 7.4.2 gives survival times (in weeks) of 20 cancer patients who received three different treatments.

Table 7.4.2. Survival times (in weeks) of three groups of cancer patients. The plus ("+") to the right of the time indicates a censored observation.

Treatment	Survival Time (in weeks)						
A	1	3	10	13^+	20	25	
B	3	8	11	20^+	25	30^+	33
C	9	15	22^+	31	35	40^+	45

The following code generates the generalized Wilcoxon, log-rank, and likelihood ratio tests using the data of Table 7.4.2.

```
data ta7_4_2;
input treatment $ weeks status @@;
datalines;
A  1 1 A  3 1 A 10 1 A 13 0 A 20 1 A 25 1
B  3 1 B  8 1 B 11 1 B 20 0 B 25 1 B 30 0 B 33 1
C  9 1 C 15 1 C 22 0 C 31 1 C 35 1 C 40 0 C 45 1
;

proc lifetest data=ta7_4_2;
time weeks*status(0);          /* Status=0 indicates censored observation */
strata treatment;             /* Identifies class variable for comparisons */
run;
```

Partial output is given below. The *p*-value using the Wilcoxon scores is 0.1441 and using the log-rank scores is 0.0721. The *p*-value for the likelihood ratio test is 0.2958.

Output – Example 7.4.2:

```
                    Test of Equality over Strata

                                        Pr >
       Test      Chi-Square   DF     Chi-Square

       Log-Rank    5.2581     2        0.0721
       Wilcoxon    3.8739     2        0.1441
       -2Log(LR)   2.4364     2        0.2958
```

8

Multivariate Permutation Tests

8.1 Tests Based on Raw Data

Suppose we have multivariate random samples from g groups. For instance, we might have random samples of heights, weights, and waist sizes of individuals from four age groups. Let $X_{ij} = \left(X_{ij1}, X_{ij2}, ..., X_{ijm} \right)$ denote the jth vector of observations from the ith group, $j = 1, ..., n_i$, $i = 1, ..., g$. We wish to test the null hypotheses that the multivariate distribution functions of the X_{ij}'s are identical against the alternative that the vectors of expected values, $E\left(X_{ij}\right)$, are not all equal. Let T denote some test statistic, to be specified momentarily, for doing this. The permutation distribution is determined by permuting the vectors X_{ij} among the groups and for each permutation computing T. The permutation p-value is the proportion of the permuted T values more extreme than the observed value.

8.1.1 Two-Sample Tests
Suppose that there are two multivariate groups to compare. For each variable of the vector $\left(X_{ij1}, X_{ij2}, ..., X_{ijm} \right)$, let T_k, $k = 1, 2..., m$, denote the two-sample t-statistic for comparing group 1 to group 2. For instance, if we have heights, weights, and waist sizes for two groups, then we would compute three t-statistics: one for comparing heights, one for comparing weights, and one for comparing waist sizes of the two groups. The statistic that SAS computes for its permutation test is the minimum of the two-sided p-values of these t-statistics. Among the set of minimum p-values obtained from permuting the data, compute the fraction that are less than or equal to the observed minimum p-value. This is the p-value of the test. Because all of the t-statistics have the same degrees of freedom, there is a one-to-one correspondence between the minimum of the two-sided p-values and the maximum of the absolute values of the t-statistics; that is, the SAS two-sample multivariate permutation test is equivalent to a test based on $T = \max\left(|T_1|, |T_2|, ..., |T_m| \right)$.

SAS Implementation
The **MULTTEST** procedure will produce a multivariate two-sample permutation test.

Example 8.1.1. An animal scientist obtained pH measurements on beef carcasses that had been subjected to one of two treatments. For each carcass, the pH measurements were made six times, with the objective being to compare the treatments across time.

Table 8.1.1. pH levels of beef carcasses taken at six times.						
Treatment	*Time 1*	*Time 2*	*Time 3*	*Time 4*	*Time 5*	*Time 6*
1	6.81	6.16	5.92	5.86	5.80	5.39
1	6.68	6.30	6.12	5.71	6.09	5.28
1	6.34	6.22	5.90	5.38	5.20	5.46
1	6.68	5.24	5.83	5.49	5.37	5.43
1	6.79	6.28	6.23	5.85	5.56	5.38
1	6.85	6.51	5.95	6.06	6.31	5.39
2	6.64	5.91	5.59	5.41	5.24	5.23
2	6.57	5.89	5.32	5.41	5.32	5.30
2	6.84	6.01	5.34	5.31	5.38	5.45
2	6.71	5.60	5.29	5.37	5.26	5.41
2	6.58	5.63	5.38	5.44	5.17	6.62
2	6.68	6.04	5.62	5.31	5.41	5.44

The following code generates a permutation test based on the resampling of *t*-statistics, using the data of Table 8.1.1.

```
data ta8_1_1;
input treat time1-time6;
datalines;
1 6.81 6.16 5.92 5.86 5.80 5.39
1 6.68 6.30 6.12 5.71 6.09 5.28
1 6.34 6.22 5.90 5.38 5.20 5.46
1 6.68 5.24 5.83 5.49 5.37 5.43
1 6.79 6.28 6.23 5.85 5.56 5.38
1 6.85 6.51 5.95 6.06 6.31 5.39
2 6.64 5.91 5.59 5.41 5.24 5.23
2 6.57 5.89 5.32 5.41 5.32 5.30
2 6.84 6.01 5.34 5.31 5.38 5.45
2 6.71 5.60 5.29 5.37 5.26 5.41
2 6.58 5.63 5.38 5.44 5.17 6.62
2 6.68 6.04 5.62 5.31 5.41 5.44
;

proc multtest data=ta8_1_1
       perm;                   /* Requests permutation p-values */
class treat;
contrast 'treat1 vs treat2' -1 1 ;   /* Contrast to be tested */
test mean(time1-time6);        /* Requests analysis on means */
run;
```

The *p*-value estimates are based on 20,000 (by default) random permutations. There is evidence that treatments differ at times 3 (*p*-value = 0.0026) and 4 (*p*-value = 0.0305) only.

***Output – Example 8.1.1*:**

```
                        The Multtest Procedure

                         Model Information

        Test for continuous variables:        Mean t-test
        Tails for continuous tests:           Two-tailed
        Strata weights:                       None
        P-value adjustment:                   Permutation
        Center continuous variables?          No
        Number of resamples:                  20000
        Seed:                                 67644
```

```
                      Contrast Coefficients

                                          treat

        Contrast                      1              2

        treat1 vs treat2             -1              1
```

Continuous Variable Tabulations

Variable	treat	NumObs	Mean	Standard Deviation
time1	1	6	6.6917	0.1858
time1	2	6	6.6700	0.0996
time2	1	6	6.1183	0.4463
time2	2	6	5.8467	0.1885
time3	1	6	5.9917	0.1514
time3	2	6	5.4233	0.1440
time4	1	6	5.7250	0.2532
time4	2	6	5.3750	0.0550
time5	1	6	5.7217	0.4266
time5	2	6	5.2967	0.0905
time6	1	6	5.3883	0.0611
time6	2	6	5.5750	0.5192

p-Values

Variable	Contrast	Raw	Permutation
time1	treat1 vs treat2	0.8064	1.0000
time2	treat1 vs treat2	0.1996	0.5817
time3	treat1 vs treat2	<.0001	0.0026
time4	treat1 vs treat2	0.0079	0.0305
time5	treat1 vs treat2	0.0381	0.1110
time6	treat1 vs treat2	0.4023	0.9364

8.1.2 *K*-sample Tests

A procedure for more than two groups may be carried out using the minimum *p*-value method. For each pair of groups, compute the *p*-values as for the two-sample case. Combine the *p*-values from all such pairs and compute the minimum of the *p*-values as the permutation statistic. For more than two groups, multiple **contrast** statements must be used. For instance, if there were three treatments instead of two for the data in Table 8.1.1, then the single **contrast** statement in the program in Example 8.1.1 would be replaced by the following three statements:

contrast 'treat1 vs treat2' 1 -1 0;
contrast 'treat1 vs treat3' 1 0 -1;
contrast 'treat2 vs treat3' 0 1 -1;

8.2 Tests Based on Ranks

8.2.1 Two-Sample and *K*-sample Tests Based on Ranks

To construct a multivariate test based on ranks, first replace the raw data with their ranks, where the ranking is done separately for each variable. Then apply the procedures outlined in Section 8.1.

SAS Implementation

The code from Example 8.1.1 can be modified to perform the analysis on the ranks of the data. First, **PROC RANK** is used to create the new data set containing the ranks of the original variables. Then the **MULTTEST** procedure is used as before on the ranks.

```
proc rank data=ta8_1_1
        out=ranks;            /* Create data set named "ranks" */
var time1-time6;
ranks rtime1-rtime6;          /* Create variables that are the ranks of the times */

proc multtest data=ranks perm;
class treat;
contrast 'treat1 vs treat2' -1 1 ;
test mean(rtime1-rtime6);
run;
```

Both *p*-values are larger than those found for the raw data. There is again strong evidence that treatments differ at time 3 (*p*-value = 0.0113), but now the evidence of a difference at time 4 is less convincing (*p*-value = 0.0521).

Output – Section 8.2.1 Example:

```
                          The Multtest Procedure

                            Model Information

            Test for continuous variables:        Mean t-test
            Tails for continuous tests:            Two-tailed
            Strata weights:                        None
            P-value adjustment:                    Permutation
            Center continuous variables?           No
            Number of resamples:                   20000
            Seed:                                  68054
```

Output – Section 8.2.1 Example (continued):

Contrast Coefficients

treat

Contrast	1	2
treat1 vs treat2	-1	1

Continuous Variable Tabulations

Variable	treat	NumObs	Mean	Standard Deviation
rtime1	1	6	7.3333	3.8816
rtime1	2	6	5.6667	3.3862
rtime2	1	6	8.5000	3.9370
rtime2	2	6	4.5000	1.8708
rtime3	1	6	9.5000	1.8708
rtime3	2	6	3.5000	1.8708
rtime4	1	6	9.0000	2.8284
rtime4	2	6	4.0000	2.3238
rtime5	1	6	8.3333	3.7238
rtime5	2	6	4.6667	2.5820
rtime6	1	6	6.0000	3.1464
rtime6	2	6	7.0000	4.2426

p-Values

Variable	Contrast	Raw	Permutation
rtime1	treat1 vs treat2	0.4464	0.9028
rtime2	treat1 vs treat2	0.0484	0.2557
rtime3	treat1 vs treat2	0.0002	0.0113
rtime4	treat1 vs treat2	0.0074	0.0521
rtime5	treat1 vs treat2	0.0756	0.3548
rtime6	treat1 vs treat2	0.6528	0.9943

8.2.2 Two-Sample Sum of Ranks Procedure

Suppose that we have two groups and that it is anticipated that group 1 will have observations that are larger than those for group 2 for all the variables. For instance, if heights, weights, and waist sizes are measured for both males and females, then we might expect that the larger observations on all three variables would occur with the males. Rank each of the variables in the combined data set, and let R_{jk} denote the rank of the jth individual for the kth variable, $j = 1, 2, ..., n_1 + n_2$,

$k = 1, 2, ..., m$. Assume that there are no tied ranks, and let $S_j = \sum_{k=1}^{m} R_{ik}$. A nonparametric procedure for comparing the two groups is a permutation test based on $\sum_{group1} S_j$, where the sum is taken over the scores for group 1. The statistic is equivalent to the sum of the Wilcoxon rank-sum statistics for group 1. The procedure is similar in philosophy to the Jonckheere-Terpstra statistic (see Section 3.4). The difference between rank averages for the two groups may be small for each variable, but the total of the ranks may show a clear difference between treatments. If there are a substantial number of ties, then an alternative procedure is to replace the R_{ij}'s with the standardized ranks

$$Z_{ij} = \frac{R_{ij} - E(R_{ij})}{\sqrt{Var(R_{ij})}}.$$

SAS Implementation

For the data in Table 8.1.1, first obtain the output data set called "ranks" as shown above. Then compute the sum of the ranks and finally use the **NPAR1WAY** procedure to compute the permutation *p*-value, using the sum of the ranks as the analysis variable. The code below shows the steps after the data set "ranks" has been created.

```
data new;              /* Create data set called "new" */
set ranks;             /* Open the data set "ranks" from previous example */
s = rtime1 + rtime2 + rtime3 + rtime4 + rtime5 + rtime6;      /* Compute sum variable */
run;

proc print data = new;            /* Print the data set "new" */
var treat rtime1-rtime6 s;
run;

proc npar1way scores=data;
class treat;
exact scores=data;
var s;                 /* Specify sum as analysis variable */
run;
```

The sum of the Wilcoxon rank-sum statistics for treatment 1 is 292 **(1)**, with exact one-sided *p*-value 0.0065 **(2)**.

Output – Section 8.2.2 Example:

Obs	treat	rtime1	rtime2	rtime3	rtime4	rtime5	rtime6	s
1	1	10	8	9	11.0	10	5.5	53.5
2	1	6	11	11	9.0	11	2.0	50.0
3	1	1	9	8	4.0	2	11.0	35.0
4	1	6	1	7	8.0	6	8.0	36.0
5	1	9	10	12	10.0	9	4.0	54.0
6	1	12	12	10	12.0	12	5.5	63.5
7	2	4	5	5	5.5	3	1.0	23.5
8	2	2	4	2	5.5	5	3.0	21.5
9	2	11	6	3	1.5	7	10.0	38.5
10	2	8	2	1	3.0	4	7.0	25.0
11	2	3	3	4	7.0	1	12.0	30.0
12	2	6	7	6	1.5	8	9.0	37.5

Output – Section 8.2.2 Example (continued):

```
                        The NPAR1WAY Procedure

                      Data Scores for Variable s
                      Classified by Variable treat

                        Sum of     Expected     Std Dev        Mean
        treat    N      Scores     Under H0     Under H0       Score
        ─────────────────────────────────────────────────────────────
        1        6      292.0      234.0        23.398718   48.666667
        2        6      176.0      234.0        23.398718   29.333333


                      Data Scores Two-Sample Test
```

(1)
```
                    Statistic (S)              292.0000

                    Normal Approximation
                    Z                            2.4788
                    One-Sided Pr >  Z            0.0066
                    Two-Sided Pr > |Z|           0.0132
```

(2)
```
                    Exact Test
                    One-Sided Pr >=  S           0.0065
                    Two-Sided Pr >= |S - Mean|   0.0130


                    Data Scores One-Way Analysis

                    Chi-Square          6.1443
                    DF                       1
                    Pr > Chi-Square     0.0132
```

9

Smoothing Methods and Robust Model Fitting

9.1 Estimating the Probability Density Function

Let $X_1, X_2, ..., X_n$ denote a random sample from a population that has a continuous probability density function $f(x)$. The *kernel method* of density estimation may be used to estimate $f(x)$. It involves taking a certain weighted average of data points near x to estimate of the density at x. Let $w(z)$ be a symmetric probability density function centered at 0. The kernel estimate of $f(x)$ is

$$\hat{f}(x) = \frac{1}{nh} \sum_{i=1}^{n} w\left(\frac{x - X_i}{h} \right)$$

The function $w(z)$ is called the *kernel,* and the quantity h is called the *bandwidth*. A commonly chosen kernel is the standard normal probability density function, but other smooth densities will give similar estimates of $f(x)$. The bandwidth plays the same role as the interval length in the histogram. If the bandwidth is too small, then the density estimate will have a choppy appearance; if it is too large, then the estimate will be overly smooth.

SAS Implementation

PROC KDE can be used for density estimation. The standard normal distribution is used for the kernel. SAS has four options for estimating the optimal bandwidth with the default being a method proposed by Sheather and Jones. See Jones et al. (1996). The value of the bandwidth may be fine-tuned by using a bandwidth multiplier.

Example 9.1.1. Table 9.1.1 gives durations (in minutes) of 107 eruptions of Old Faithful Geyser.

Table 9.1.1. Durations (in minutes) of eruptions of Old Faithful geyser										
1.7	1.8	2.0	2.9	3.5	3.8	4.0	4.1	4.3	4.5	4.6
1.7	1.8	2.0	2.9	3.6	3.8	4.0	4.1	4.3	4.5	4.6
1.7	1.8	2.0	3.1	3.6	3.8	4.0	4.1	4.3	4.5	4.7
1.7	1.8	2.0	3.2	3.7	3.8	4.0	4.1	4.3	4.6	4.7
1.7	1.9	2.0	3.3	3.7	3.8	4.0	4.1	4.4	4.6	4.7
1.8	1.9	2.3	3.4	3.7	3.9	4.0	4.1	4.4	4.6	4.8
1.8	1.9	2.3	3.4	3.7	3.9	4.0	4.2	4.4	4.6	4.9
1.8	1.9	2.3	3.5	3.7	3.9	4.1	4.2	4.4	4.6	
1.8	1.9	2.3	3.5	3.7	3.9	4.1	4.3	4.4	4.6	
1.8	1.9	2.5	3.5	3.7	4.0	4.1	4.3	4.5	4.6	

The following code generates density estimates using the kernel method and also uses **PROC GPLOT** to produce a plot of the estimated density, using the data of Table 9.1.1.

The **bwm=** *bandwidth multiplier* option can be used to alter the bandwidth to obtain more or less smoothing, with larger multipliers producing smoother curves.

```
data ta9_1_1;
input minutes@@;
datalines;
1.7 1.9 3.4 3.8 4.1 4.5
1.7 1.9 3.5 3.9 4.1 4.5
1.7 2.0 3.5 3.9 4.1 4.5
1.7 2.0 3.5 3.9 4.1 4.6
1.7 2.0 3.5 3.9 4.2 4.6
1.8 2.0 3.6 4.0 4.2 4.6
1.8 2.0 3.6 4.0 4.3 4.6
1.8 2.3 3.7 4.0 4.3 4.6
1.8 2.3 3.7 4.0 4.3 4.6
1.8 2.3 3.7 4.0 4.3 4.6
1.8 2.3 3.7 4.0 4.3 4.6
1.8 2.5 3.7 4.0 4.3 4.6
1.8 2.9 3.7 4.0 4.4 4.7
1.8 2.9 3.7 4.1 4.4 4.7
1.9 3.1 3.8 4.1 4.4 4.7
1.9 3.2 3.8 4.1 4.4 4.8
1.9 3.3 3.8 4.1 4.4 4.9
1.9 3.4 3.8 4.1 4.5
;
```

```
proc kde data=ta9_1_1
        bwm=1         /* Sets the bandwidth */
        gridl=1.7     /* Specifies the lower grid limit for the kernel density estimate */
        gridu=4.9     /* Specifies the lower grid limit for the kernel density estimate */
        out=new;      /* Names new data set containing information to plot the density */
var minutes;          /* Specifies analysis variable */
run;

proc gplot data=new;  /* Plots estimated density */
plot density*minutes;
run;
```

Output – Example 9.1.1:

```
                     The KDE Procedure

                          Inputs

          Data Set                       WORK.EX10_1_1
          Number of Observations Used    107
          Variable                       minutes
          Bandwidth Method               Sheather-Jones
                                         Plug In
```

Output – Example 9.1.1 (continued):

```
                          Controls

                                       minutes

        Grid Points                      401
        Lower Grid Limit                 1.7
        Upper Grid Limit                 4.9
        Bandwidth Multiplier               1


                          Statistics

                                       minutes

        Mean                            3.46
        Variance                        1.07
        Standard Deviation              1.04
        Range                           3.20
        Interquartile Range             2.00
        Bandwidth                       0.19


                         Percentiles

                                       minutes

              0.5                        1.70
              1.0                        1.70
              2.5                        1.70
              5.0                        1.80
             10.0                        1.80
             25.0                        2.30
             50.0                        3.80
             75.0                        4.30
             90.0                        4.60
             95.0                        4.60
             97.5                        4.70
             99.0                        4.80
             99.5                        4.90


                            Levels

        Percent    Density    Lower    Upper

              1     0.06265     1.70     4.90
              5     0.1298      1.70     4.90
             10     0.1607      1.70     4.87
             50     0.4284      1.80     4.59
             90     0.5492      3.97     4.10
             95     0.5535      4.00     4.07
             99     0.5535      4.00     4.07
            100     0.5550      4.04     4.04
```

Output – Example 9.1.1 (continued):

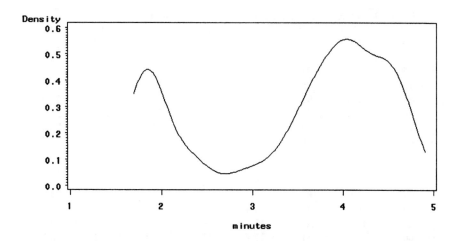

9.2   Nonparametric Curve Smoothing

To estimate a regression function $y = \lambda(x)$ at $x = x_0$ when the functional form is not known, the *loess method* may be used. It approximates $\lambda(x)$ "locally" by a linear or quadratic function $l(x)$. To fit $l(x)$ locally, the fraction of observations nearest to x_0 to be used in fitting the equation is specified. This set of nearest points is denoted as $N(x_0)$. The function $l(x)$ is fitted to the pairs $(x_i, y_i), x_i \in N(x_0)$, using weighted least squares. The estimate of $y = \lambda(x)$ at $x = x_0$ is $\hat{y} = l(x_0)$. The value of fraction used in determining $N(x_0)$ is called the *smoothing parameter*, and the larger its value, the smoother the curve. The procedure may be extended to multiple predictor variables where the locally fitted models are either multivariable linear or quadratic. The quadratic models involve linear, squared, and cross-product terms of the predictor variables.

SAS Implementation
PROC LOESS can implement the loess method of curve smoothing. SAS uses the local linear fit as the default, but the quadratic option is also available. For the fit at $x = x_0$, the weight associated with observation x_i is proportional to $w_i = \left(1 - d_i^3\right)^3$, where d_i is the distance from x_i to x_0 divided by the maximum distance of points within $N(x_0)$ from x_0. SAS allows the smoothing parameter to be bigger than 1, and in this case d_i is divided by $s^{1/p}$, where s is the fraction and p is the number of predictors in the equation.

Example 9.2.1. The following table gives the horsepower (HP) and miles per gallon (MPG) of 82 automobiles.

Table 9.2.1. Horsepower and miles per gallon of 82 automobiles.

HP	MPG	HP	MPG	HP	MPG	HP	MPG
49	65.4	74	407	95	32.2	140	25.3
55	56.0	95	40.0	102	32.2	140	23.9
55	55.9	81	39.3	95	32.2	150	23.6
70	49.0	95	38.8	93	31.5	165	23.6
53	46.5	92	38.4	100	31.5	165	23.6
70	46.2	92	38.4	100	31.4	165	23.6
55	45.4	92	38.4	98	31.4	165	23.6
62	59.2	90	29.5	130	31.2	245	23.5
62	53.3	52	46.9	115	33.7	280	23.4
80	43.4	103	36.3	115	32.6	162	23.4
73	41.1	84	36.1	115	31.3	162	23.1
92	40.9	84	36.1	115	31.3	140	22.9
92	40.9	102	35.4	180	30.4	140	22.9
73	40.4	102	35.3	160	28.9	175	19.5
66	39.6	81	35.1	130	28.0	322	18.1
73	39.3	90	35.1	96	28.0	238	17.2
78	38.9	90	35.0	115	28.0	263	17.0
92	38.8	102	33.2	100	28.0	295	16.7
78	38.2	102	32.9	100	28.0	236	13.2
90	42.2	130	32.3	145	27.7		
92	40.9	95	32.2	120	25.6		

The following code generates the loess smoother of HP versus MPG for the data of Table 9.2.1 and provides a plot of the smoothed density estimates.

```
data ta9_2_1;
input HP MPG@@;
datalines;
49 65.4 74 40.7 95 32.2 140 25.3
55 56.0 95 40.0 102 32.2 140 23.9
55 55.9 81 39.3 95 32.2 150 23.6
70 49.0 95 38.8 93 31.5 165 23.6
53 46.5 92 38.4 100 31.5 165 23.6
70 46.2 92 38.4 100 31.4 165 23.6
55 45.4 92 38.4 98 3.4 165 23.6
62 59.2 90 29.5 130 31.2 245 23.5
62 53.3 52 46.9 115 33.7 280 23.4
80 43.4 103 36.3 115 32.6 162 23.4
73 41.1 84 36.1 115 31.3 162 23.1
92 40.9 84 36.1 115 31.3 140 22.9
92 40.9 102 35.4 180 30.4 140 22.9
73 40.4 102 35.3 160 28.9 175 19.5
66 39.6 81 35.1 130 28.0 322 18.1
73 39.3 90 35.1 96 28.0 238 17.2
78 38.9 90 35.0 115 28.0 263 17.0
92 38.8 102 33.2 100 28.0 385 16.7
```

```
78 38.2 102 32.9 100 28.0 236 13.2
90 42.2 130 32.3 145 27.7
92 40.9 95 32.2 120 25.6
;

proc loess data=ta9_2_1;
model MPG=HP / smooth=0.5;        /* Specifies smoothing parameter of 0.5 */
ods output OutputStatistics=Results;    /* Creates output data set */
run;

proc sort data=Results out=sorted;       /* Sorts data for proper plotting */
by hp;
run;

symbol1 color=black v=plus;
symbol2 color=black v=none i=join;
proc gplot data=sorted;
plot DepVar*HP Pred*HP/overlay;    /* MPG named "DepVar" in output data set */
run;
quit;
```

***Output – Example 9.2.1*:**

```
                    The LOESS Procedure
                  Smoothing Parameter: 0.5
                  Dependent Variable: MPG

                        Fit Summary

         Fit Method                      kd Tree
         Blending                        Linear
         Number of Observations          82
         Number of Fitting Points        17
         kd Tree Bucket Size             8
         Degree of Local Polynomials     1
         Smoothing Parameter      ·      0.50000
         Points in Local Neighborhood    41
         Residual Sum of Squares         2132.24702
```

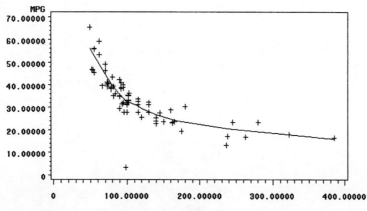

9.3   Robust Regression

When fitting a multiple regression equation to data by the method of least squares, a few observations that do not follow the pattern of the others may have a large influence on the values of the estimated coefficients. The objective in robust regression is to fit equations in ways that are not so sensitive to these influential observations.

9.3.1. Least Absolute Values Regression

In least absolute values regression, the coefficients of the regression equation are chosen to minimize the sum of the absolute residuals $\sum |observed - predicted|$.

SAS Implementation

There are two different ways to obtain regression results using least absolute values. One is to use **PROC NLP** in the SAS/OR software component. Another is to use the LAV function in **PROC IML**. The first set of code illustrates the use of **PROC NLP**, using the data set created in Example 9.2.1. Several different optimization algorithms are available, with the default being a quasi-Newton algorithm, which is specified in the **tech=quanew** option in the program below. The **min absval** option specifies that the function defined as "absval" be minimized.

```
proc nlp data=ta9_2_1
tech=quanew;          /* Specifies optimization technique */
min absval;           /* Specifies function to be minimized */
parms b1=1, b0=1;         /* Specifies initial parameter estimates */
absval=abs(mpg - (b0 + b1*hp));   /* Defines function to be minimized */
run;
```

The majority of the output produced gives iteration details of the optimization technique. The last page of the output is given below. The parameter estimates are –0.16 for the slope and 51.04 for the intercept. The minimum value obtained for the sum of the absolute values of the residuals was 408.81. For comparison, the least squares estimates are –0.12 and 48.1 for the slope and intercept, respectively.

Output – Section 9.3.1 NLP Example:

```
                         L1-Norm Parameter Estimates
                      Minimize Absolute Value of Residual
                     Using Least_Squares Estimates to Start

                       PROC NLP: Nonlinear Minimization

                            Optimization Results
                            Parameter Estimates
                                                         Gradient
                                                         Objective
                N Parameter          Estimate            Function

                1 b1               -0.160330          -449.000000
                2 b0               51.037945            -2.000000

             Value of Objective Function = 408.8124072
```

Another way to obtain the above results is by using **PROC IML**. The following code calculates the same estimates as above. The code assumes that data file Work.T9_2_1 has already been created. The **lav** function takes the input matrices and returns the results. The x and y variables are first read into vectors, and then a column of 1's is appended to the x vector to create the X matrix. In the function call, rc is a scalar that holds the result code for the convergence of the optimization algorithm ($rc = 0$ indicates the algorithm has converged), xr contains the parameter estimates, and x and y are the input matrix and vectors, respectively.

```
proc iml;
use ta9_2_1;
read all var {HP} into x;      /* Create vector for independent variable */
read all var {MPG} into y;     /* Create vector for dependent variable */
m=nrow(x);                     /* Determine number of observations */
x=j(m,1,1.)||x;                /* Create design matrix */
call lav(rc,xr,x,y);           /* Function call */
print rc xr;                   /* Print results */
quit;
```

The algorithm obtains the same estimates of slope (–0.16) and intercept (51.04) as were obtained by **PROC NLP** in the previous example.

Output – Section 9.3.1 IML Example:

```
                              LAV (L1) Estimation

                            Start with LS Solution
                   Start Iter: gamma=32.544976358 ActEqn=82
Iter    N Huber    Act Eqn     Rank     Gamma       L1(x)       F(Gamma)

  1        1         10         2       0.7833    417.724797   387.207577
  1        2         11         2       0.7833    409.507794   378.384664
  1        3          8         2       0.7833    409.192626   378.144968
  2        4          7         2       0.5373    409.162739   387.593011
  3        5          6         2       0.4549    409.119034   390.803117
  4        6          5         2       0.3558    409.034032   394.692682
  5        7          4         2       0.2373    409.002395   399.380427
  6        8          3         2       0.0724    408.819422   405.916248
  7        9          2         2       0.0132    408.816744   408.274879
  7        9          2         2       0.0000    408.812397   408.274879

                            Algorithm converged
                 Objective Function L1(x)= 408.81239669
                    Number Search Directions=       17
                    Number Refactorizations =        3
                    Total Execution Time=      0.0000


                   RC          XR

               0 51.038017 -0.160331
```

9.3.2. *M*-Estimation

In *M-estimation*, the coefficients are chosen to minimize functions of the standardized residuals of the form $\sum \rho(\text{standarized residual})$, where $\rho(x)$ is an appropriately chosen symmetric function with a unique minimum at $x = 0$. One such function is the *Tukey bi-square* function defined by

$$\rho(x) = \left(\frac{x}{d}\right)^6 - 3\left(\frac{x}{d}\right)^4 + 3\left(\frac{x}{d}\right)^2, |x| \le d,$$

and $\rho(x) = 1$ otherwise.

SAS Implementation

SAS Version 8 and below will not perform *M*-Estimation. Version 9, however, contains an experimental procedure call **PROC ROBUSTREG** that will perform *M*-Estimation, as well as some other robust methods. *M*-Estimation is the default method, and the default weight function is the Tukey bi-square function with constant $d = 2.5$. Other values of d can be specified by the user. See SAS documentation for more information on other weight functions.

The following code illustrates using **PROC ROBUSTREG** to perform *M*-Estimation, again using the data file created in Exercise 9.2.1. Parameter estimates and large sample confidence intervals and *p*-values are given. The estimates are –0.13 for the slope and 48.47 for the intercept.

```
proc robustreg data=ta9_2_1
      method=m           /* Specifies M-estimation */
      (wf=bisquare);     /* Specifies Tukey bisquare weight function */
model mpg=hp;
run;
```

***Output – Section 9.3.2*:**

The ROBUSTREG Procedure

Model Information

Data Set	WORK.TA9_2_1
Dependent Variable	MPG
Number of Independent Variables	1
Number of Observations	82
Method	M Estimation

Number of Observations Read	82
Number of Observations Used	82

Summary Statistics

Variable	Q1	Median	Q3	Mean	Standard Deviation	MAD
HP	84.0000	99.0000	140.0	118.2	61.0321	31.1346
MPG	25.6000	32.4500	39.3000	33.4402	10.5499	10.1558

Parameter Estimates

Parameter	DF	Estimate	Standard Error	95% Confidence Limits		Chi-Square	Pr > ChiSq
Intercept	1	48.4705	1.3520	45.8207	51.1202	1285.36	<.0001
HP	1	-0.1321	0.0102	-0.1520	-0.1121	168.53	<.0001
Scale	1	5.5197					

Diagnostics Summary

Observation Type	Proportion	Cutoff
Outlier	0.0488	3.0000

Goodness-of-Fit

Statistic	Value
R-Square	0.4848
AICR	91.1861
BICR	97.1649
Deviance	2691.765

Example 9.3.1. This example illustrates the methods from this section applied to a model with more than one predictor. The following table gives collar size (in inches), shoe size, and weight (in pounds) for 24 college-aged men.

Table 9.3.1. Collar size (in inches), shoe size, and weight for 24 college-aged men.							
ID	CS	SS	WT	ID	CS	SS	WT
1	14.5	9.5	140	13	15.5	11	180
2	15.5	9.5	155	14	15.5	11	175
3	15.5	10.5	153	15	15.5	10.5	155
4	15	10.5	150	16	15.5	8.5	150
5	16.5	11	180	17	15	8.5	180
6	16.5	8.5	160	18	15.5	10	160
7	15.5	8.5	155	19	15	9	145
8	14.5	9.5	145	20	16	12	190
9	15	10	163	21	16.5	13	228
10	15	9	150	22	15	8.5	150
11	15	8.5	140	23	15	11	165
12	15.5	9.5	170	24	15	9	145

The following code produces least absolute value regression estimates using **PROC NLP** and **PROC IML**, *M*-Estimation using **PROC ROBUSTREG**, and least squares estimates using **PROC REG**.

```
data ta9_3_1;
input cs ss wt;
datalines;
14.5   9.5    140
15.5   9.5    155
15.5   10.5   153
15     10.5   150
16.5   11     180
16.5   8.5    160
15.5   8.5    155
14.5   9.5    145
15     10     163
15     9      150
15     8.5    140
15.5   9.5    170
15.5   11     180
15.5   11     175
15.5   10.5   155
15.5   8.5    150
15     8.5    180
15.5   10     160
15     9      145
16     12     190
16.5   13     228
15     8.5    150
15     11     165
15     9      145
;
```

```
proc nlp data= ta9_3_1
       tech=quanew;          /* Specifies optimization technique */
       min absval;           /* Specifies function to be minimized */
parms b2=1, b1=1, b0=1;      /* Specifies initial parameter estimates */
absval=abs(wt - (b0 + b1*cs+b2*ss));      /* Defines function to be minimized */
title 'PROC NLP Results';
run;

title 'PROC IML Results';
proc iml;
use ta9_3_1;
read all var {cs ss} into x;   /* Create vector for independent variable */
read all var {wt} into y;      /* Create vector for dependent variable */
m=nrow(x);                     /* Determine number of observations */
x=j(m,1,1.)||x;                /* Create design matrix */
call lav(rc,xr,x,y);           /* Function call */
print rc xr;                   /* Print results */
quit;

proc robustreg data=ta9_3_1
       method=m;    /* Specifies M-estimation */
model wt=cs ss;
title 'PROC ROBUSTREG Results';
run;

proc reg data=ta9_3_1;
model wt=cs ss;
run;
```

A summary of the estimates provided by each procedure is given in Table 9.3.2. Edited output produced by the procedures is given below.

Parameter	Parameter estimates			
	NLP	**IML**	**ROBUSTREG**	**REG**
Intercept	−149.05	−149.05	−127.46	−156.57
β_1	13.33	13.33	13.27	15.07
β_2	10.48	10.48	8.40	8.79

Output - Example 9.3.1:

```
                           PROC NLP Results
                     PROC NLP: Nonlinear Minimization

                          Optimization Results
                          Parameter Estimates

                                                    Gradient
                                                    Objective
                    N Parameter        Estimate     Function

                    1 b2              10.476190     22.500000
                    2 b1              13.333331     32.000000
                    3 b0            -149.047586      2.000000

          Value of Objective Function = 172.28571491
```

Output - Example 9.3.1 (continued)**:**

PROC IML Results

LAV (L1) Estimation
Start with LS Solution
Start Iter: gamma=35.726249769 ActEqn=24

Iter	N Huber	Act Eqn	Rank	Gamma	L1(x)	F(Gamma)
1	1	5	3	1.3299	179.736053	164.574488
1	2	2	3	1.3299	174.172316	158.914138
1	3	7	2	1.3299	173.111990	157.598898
2	4	6	3	0.9163	172.859025	161.969912
3	5	5	3	0.7647	172.722436	163.655687
4	6	5	3	0.7143	172.693878	164.224490
4	6	3	3	0.0000	172.285714	164.224490

Algorithm converged
Objective Function L1(x)= 172.28571429
Number Search Directions=   11
Number Refactorizations =   4
Total Execution Time=   0.0100

RC		XR	
0	-149.0476	13.333333	10.47619

PROC ROBUSTREG Results

The ROBUSTREG Procedure

Model Information

Data Set	WORK.TA9_3_1
Dependent Variable	wt
Number of Independent Variables	2
Number of Observations	24
Method	M Estimation

Number of Observations Read   24
Number of Observations Used   24

Summary Statistics

Variable	Q1	Median	Q3	Mean	Standard Deviation	MAD
cs	15.0000	15.5000	15.5000	15.3750	0.5566	0.7413
ss	8.7500	9.5000	10.7500	9.8542	1.2290	1.4826
wt	150.0	155.0	172.5	161.8	19.7829	14.8260

Parameter Estimates

Parameter	DF	Estimate	Standard Error	95% Confidence Limits		Chi-Square	Pr > ChiSq
Intercept	1	-127.463	47.3375	-220.243	-34.6831	7.25	0.0071
cs	1	13.2690	3.3904	6.6239	19.9141	15.32	<.0001
ss	1	8.3979	1.5354	5.3885	11.4073	29.91	<.0001
Scale	1	6.5235					

Output - Example 9.3.1 (continued):

```
                    Least Squares Estimates

                    The REG Procedure
                     Model: MODEL1
                  Dependent Variable: wt

        Number of Observations Read        24
        Number of Observations Used        24


                   Analysis of Variance

                          Sum of        Mean
    Source           DF   Squares       Square    F Value    Pr > F

    Model             2   6176.90811    3088.45405    22.96    <.0001
    Error            21   2824.42522    134.49644
    Corrected Total  23   9001.33333


            Root MSE            11.59726    R-Square    0.6862
            Dependent Mean     161.83333    Adj R-Sq    0.6563
            Coeff Var            7.16618


                   Parameter Estimates

                     Parameter    Standard
       Variable   DF   Estimate       Error    t Value    Pr > |t|

       Intercept   1   -156.57217    67.88615     -2.31    0.0314
       cs          1     15.07381     4.86216      3.10    0.0054
       ss          1      8.79279     2.20196      3.99    0.0007
```

9.3.3. Other Robust Methods

The **ROBUSTREG** procedure will also perform Least Trimmed Squares (LTS) estimation (see Rousseeuw, 1984), *S*-Estimation (see Rousseeuw & Van Driessen, 2000), and *MM*-Estimation (see Yohai, 1987).

PROC IML has a function that will perform LTS, and in addition has functions that can calculate least median of squares (LMS) estimation, minimum volume ellipsoid (MVE) estimation, and minimum covariance determinant (MCD) estimation. See Rousseeuw (1984) and Rousseeuw and Leroy (1987) for more details.

References

Agresti, A. (2002). *Categorical Data Analysis*, 2nd ed. John Wiley & Sons, New York.

Blair, R. C., and Higgins, J. J. (1980). A Comparison of the Power of Wilcoxon's Rank-Sum Test to that of Student's T Statistic Under Various Non-Normal Distributions. *Journal of Educational Statistics*, 5(4), 309–335.

Brown, M. B., and Benedetti, J. K. (1977). Sampling Behavior of Tests for Correlation in Two-Way Contingency Tables. *Journal of the American Statistical Association*, 72, 309–315.

Cochran, W. G. (1963). *Sampling Techniques*, 2nd ed. John Wiley & Sons, New York.

Conover, W. J. (1999). *Practical Nonparametric Statistics*, 3rd ed. John Wiley & Sons, New York.

Good, P. (2000). *Permutation Tests. A Practical Guide to Resampling Methods for Testing Hypotheses*, 2nd ed. Springer, New York.

Hettmansperger, T. P., and McKean, J. W. (1998). *Robust Nonparametric Statistical Methods,* Arnold, London.

Higgins, J. J. (2004). *An Introduction to Modern Nonparametric Statistics,* Brooks/Cole—Thompson Learning, Pacific Grove, CA.

Hollander, M., and Wolfe, D. (1999). *Nonparametric Statistical Methods*, 2nd ed. John Wiley & Sons, New York.

Jones, M. C., Marron, J. S., and Sheather, S. J. (1996). A Brief Survey of Bandwidth Selection for Density Estimation. *Journal of the American Statistical Association*, 91, 401–407.

Lawless, J. F. (2002). *Statistical Models and Methods for Lifetime Data,* 2nd ed. John Wiley & Sons, New York.

Lee, E. T. (2003). *Statistical Methods for Survival Data Analysis*, 3rd ed. John Wiley & Sons, New York.

Lehmann, E. L. (1975). *Nonparametrics: Statistical Methods Based on Ranks*, Holden-Day, San Francisco.

Pitman, E. J. G. (1937). Significance Tests Which May be Applied to Samples from Any Population. *Journal of the Royal Statistical Society*, Series B, 4, 119–130.

Rousseeuw, P. J. (1984). Least Median of Squares Regression. *Journal of the American Statistical Association*, 79, 871–880.

Rousseeuw, P. J., and Leroy, A. M. (1987). *Robust Regression and Outlier Detection*, John Wiley & Sons, Inc, New York.

Rousseeuw, P. J., and Van Driessen, K. (2000). An Algorithm for Positive-Breakdown Regression Based on Concentration Steps. *Data Analysis: Scientific Modeling and Practical Application*, ed. W. Gaul, O. Opitz, and M. Schader. New York: Springer-Verlag, 335–346.

SAS Institute Inc. (1999). *SAS OnlineDoc®, Version 8*. SAS Institute Inc., Cary, NC.

SAS Institute Inc. (2004). *SAS OnlineDoc® 9.1.2*. SAS Institute Inc., Cary, NC.

Westfall, P., Tobias, R., Rom, D., Wilfinger, R., and Hochberg, Y. (1999). *Multiple Comparisons and Multiple Tests Using the SAS® System*. SAS Institute Inc., Cary, NC.

Yohai V. J. (1987). High Breakdown Point and High Efficiency Robust Estimates for Regression. *Annals of Statistics*, 15, 642–656.

Index